Proteomics

A COLD SPRING HARBOR
LABORATORY COURSE MANUAL

ALSO FROM COLD SPRING HARBOR LABORATORY PRESS

RELATED LABORATORY MANUALS

Basic Methods in Protein Purification and Analysis: A Laboratory Manual

Methods in Yeast Genetics: A Cold Spring Harbor Laboratory Course Manual

Protein–Protein Interactions: A Molecular Cloning Manual, Second Edition

Proteins and Proteomics: A Laboratory Manual

OTHER RELATED TITLES

At the Bench: A Laboratory Navigator, Updated Edition

At the Helm: A Laboratory Navigator

Bioinformatics: Sequence and Genome Analysis, Second Edition

Lab Math: A Handbook of Measurements, Calculations, and Other Quantitative Skills for Use at the Bench

Lab Ref, Volumes 1 and 2: A Handbook of Recipes, Reagents, and Other Reference Tools for Use at the Bench

Experimental Design for Biologists

RELATED WEBSITE

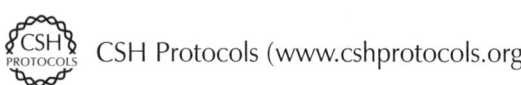 CSH Protocols (www.cshprotocols.org)

Proteomics

A COLD SPRING HARBOR LABORATORY COURSE MANUAL

Andrew J. Link

Vanderbilt University School of Medicine

Joshua LaBaer

*Harvard University School of Medicine and
Harvard Institute of Proteomics*

COLD SPRING HARBOR LABORATORY PRESS
Cold Spring Harbor, New York • www.cshlpress.com

Proteomics
A COLD SPRING HARBOR LABORATORY COURSE MANUAL

All rights reserved.
© 2009 by Cold Spring Harbor Laboratory Press, Cold Spring Harbor, New York
Printed in the United States of America

Publisher	John Inglis
Acquisition Editor	Alexander Gann
Book Development, Marketing and Sales Director	Jan Argentine
Developmental Editor	Michael Zierler
Project Coordinator	Maryliz Dickerson
Permissions Coordinator	Carol Brown
Production Editors	Kathleen Bubbeo and Kaaren Kockenmeister
Desktop Editor	Susan Schaefer
Production Manager	Denise Weiss
Book Marketing Manager	Ingrid Benirschke
Sales Account Manager	Elizabeth Powers
Cover Design	Ed Atkeson

Front cover artwork (*for paperback edition only*): DNA detection (PicoGreen). Images courtesy of Sanjeeva Srivastava and Niroshan Ramachandran, Harvard Medical School.

Library of Congress Cataloging-in-Publication Data

Link, Andrew J.
　Proteomics : a Cold Spring Harbor laboratory course manual / Andrew Link, Joshua LaBaer.
　　　p. cm.
　　Includes index.
　　ISBN 978-0-87969-787-7 (pbk. : alk. paper) -- ISBN 978-0-87969-793-8 (hardcover : alk. paper)
　1. Proteomics--Laboratory manuals. I. LaBaer, Joshua. II. Title.

QP551.L53 2009
572'.6078--dc22

2008041374

10 9 8 7 6 5 4 3 2 1

Students and researchers using the procedures in this manual do so at their own risk. Cold Spring Harbor Laboratory makes no representations or warranties with respect to the material set forth in this manual and has no liability in connection with the use of these materials. All registered trademarks, trade names, and brand names mentioned in this book are the property of the respective owners. Readers should please consult individual manufacturers and other resources for current and specific product information.

With the exception of those suppliers listed in the text with their addresses, all suppliers mentioned in this manual can be found on the BioSupplyNet website at http://www.biosupplynet.com.

All World Wide Web addresses are accurate to the best of our knowledge at the time of printing.

Procedures for the humane treatment of animals must be observed at all times. Check with the local animal facility for guidelines.

Certain experimental procedures in this manual may be the subject of national or local legislation or agency restrictions. Users of this manual are responsible for obtaining the relevant permissions, certificates, or licenses in these cases. Neither the authors of this manual nor Cold Spring Harbor Laboratory assume any responsibility for failure of a user to do so.

The polymerase chain reaction (PCR) process is covered by certain patent and proprietary rights. Users of this manual are responsible for obtaining any licenses necessary to practice PCR or to commercialize the results of such use. COLD SPRING HARBOR LABORATORY MAKES NO REPRESENTATION THAT USE OF THE INFORMATION IN THIS MANUAL WILL NOT INFRINGE ANY PATENT OR OTHER PROPRIETARY RIGHT.

Authorization to photocopy items for internal or personal use, or the internal or personal use of specific clients, is granted by Cold Spring Harbor Laboratory Press, provided that the appropriate fee is paid directly to the Copyright Clearance Center (CCC). Write or call CCC at 222 Rosewood Drive, Danvers, MA 01923 (508-750-8400) for information about fees and regulations. Prior to photocopying items for educational classroom use, contact CCC at the above address. Additional information on CCC can be obtained at CCC Online at http://www.copyright.com/.

All Cold Spring Harbor Laboratory Press publications may be ordered directly from Cold Spring Harbor Laboratory Press, 500 Sunnyside Blvd., Woodbury, N.Y. 11797-2924. Phone: 1-800-843-4388 in Continental U.S. and Canada. All other locations: (516) 422-4100. FAX: (516) 422-4097. E-mail: cshpress@cshl.edu. For a complete catalog of all Cold Spring Harbor Laboratory Press publications, visit our World Wide Web site at http://www.cshlpress.com/.

Contents

Preface, vii
Introduction, 1

EXPERIMENTS

1. Analysis of Whole-cell Lysates by Two-dimensional Gel Electrophoresis and MALDI Mass Spectrometry, 13

2. Purification of Protein Complexes for Mass Spectrometry Analysis, 37

3. Qualitative and Quantitative Measurement of Peptides with MALDI TOF/TOF Mass Spectrometry, 55

4. Analysis of Protein Complexes: High-sensitivity Liquid Chromatography Coupled with Tandem Mass Spectrometry, 69

5. Phosphopeptide Analysis Using IMAC and Mass Spectrometry, 83

6. Multidimensional Protein Identification Technology (MudPIT) Analysis of Whole-cell Lysates, 95

7. Quantitative Mass Spectrometry Analysis of Whole-cell Extracts (iTRAQ), 103

8. Analysis and Validation of Tandem Mass Spectra, 117

9. High-throughput Cloning of ORFs: Assembling Large Sets of Expression Constructs, 143

10. Construction of Protein Microarrays: Nucleic Acid Programmable Protein Array (NAPPA), 153

11. Using the Nucleic Acid Programmable Protein Array (NAPPA) for Identifying Protein–Protein Interactions, 175

APPENDICES

1. Setup and Demonstration of a Nanoelectrospray Ionization (nanoESI) Source and Tandem Mass Spectrometry (MS/MS), 183

2. Solution Protein Digest, 189

3. In-gel Trypsin Digest of Gel-fractionated Proteins, 193

4. Trichloroacetic Acid (TCA) Precipitation of Proteins, 197

5. Monoisotopic and Immonium Ion Masses of Amino Acids, 199

6. Dipeptide Masses of Amino Acids, 201

7. LTQ Instrument Methods, 203

8. Off-line Desalting of Peptide Mixtures, 211

9. Preparing Competent Cells, 215

10. DNA Quantification, 217

11. Cautions, 219

Index, 225

Preface

EVEN AS THE DNA SEQUENCING EXPERTS were racing to complete the human genome sequence and excitement in the press had grown to a fever pitch, many of us could not tear ourselves away from the most interesting macromolecules in biology—the proteins. Sure, sequencing DNA is much easier than identifying, sequencing, or studying proteins. Splice variations and the large number of possible posttranslational processing steps, from adducts to selective cleavage, create a staggering number of potential protein species. Proteins are far more complex than any other macromolecules; the unique chemistry of each protein makes it nearly impossible to develop methods that suit all proteins. But, don't the best things in life require the most effort?

Life happens at the level of proteins. Proteins provide the verbs to biology. They build, process, activate, and inactivate; they polymerize, repair, support, modify, degrade, fold, migrate, and transport; they shorten, signal, cleave, inhibit, digest, fluoresce, induce, excise, carry, and repress; they bind, transfer, translocate, amplify, proofread, regulate, and perform countless more activities. When things go wrong, proteins are inevitably at the center of the problem. Most disease is the result of protein dysfunction and virtually all drugs act by modifying protein function.

The execution of the genome sequencing projects was a watershed in the study of proteins and their functions. Genome sequence information has produced the predicted amino acid sequences of most proteins, leading to the development of informatics tools that are now used routinely in the identification of proteins by mass spectrometry. This information also identified gene sequences that can be cloned to produce recombinant protein. And, in a larger sense, the sequencing projects introduced a culture to biology that was already common in the physical sciences: a reliance on multidisciplinary team efforts. Thus began a new approach to the study of proteins, termed proteomics, which adapted the large-scale strategies and multidisciplinary approaches of the genome projects to the study of proteins and their roles in biology.

Virtually all types of experiments with DNA are fundamentally based on variations on the same basic chemistry—the pairing of the nucleotide bases by hydrogen bonds. In contrast, every protein displays a unique chemistry and structure that makes universal methods impossible. Necessarily, tools for proteomics are both more numerous and more complex than those used for studying nucleic acids and we are still learning how to apply them at scale.

Interest in the proteomics approach to studying proteins exploded at the beginning of this century. This rapidly growing interest suggested the need for a course that taught these methods. In particular, students, postdoctoral fellows, technicians, lab managers, and principal investigators all wanted to use proteomics in their research, but most did not know how. Even when they could find someone else to collect the data on their samples, they were concerned that they did not understand the methods well enough to interpret the data. What could be trusted and what was suspect?

To address this need, the two of us, along with Philip Andrews at the University of Michigan, started the Proteomics course at Cold Spring Harbor Laboratory in 2002, which is an intensive, hands-on introduction to the field. The fundamental approach to the course is to annually select and teach 16 students about proteomics through performing actual procedures and interpreting the data. As evidenced by this book, the course covers a broad range of topics in proteomics that includes various protein preparation and separation methods, common methods for performing mass spectrometry, high-throughput cloning of coding sequences for protein expression, production and use of protein microarrays, and analysis methods for examining these data. The course teaches some of the most advanced methods in proteomics, including many new methods prior to their publication. Since the course was first started, it has met with profound enthusiasm from students, annually receiving five times as many applicants as the course can handle.

This laboratory manual contains step-by-step protocols for conducting most of the experiments that have been taught over the past 6 years. The instructions are detailed, including notes about technique and specific product numbers in order to help students reproduce these methods in their own labs. In a fast-moving field like proteomics, some of these methods will be supplanted by newer ones in the coming years. The methods taught in our course will continue to evolve as new methods and new approaches develop for studying proteins. Still, all of the methods in this course manual are time-tested and robust. We intend to update this manual periodically as the content of the course changes.

This work would not be possible without the help of our many outstanding teaching assistants who helped to write these protocols, and the many members of our labs and of the mass spectrometry, analytical chemistry, bioinformatics, and biological community who helped to develop them. We are indebted to them all.

JOSHUA LaBAER
ANDREW J. LINK

Introduction

Proteomics has emerged as a powerful approach for studying biological processes and directly profiling changes in cells and tissues. In medicine, proteomics is a major research tool to look for new biomarkers for human diseases and to discover new targets for drug therapies (Omenn 2006). Most current drug therapies to treat human diseases target proteins.

The Cold Spring Harbor Laboratory Proteomics course and this manual were designed to train students, postdoctoral fellows, and senior investigators in the advanced techniques and applications of proteomics in biomedical research. In this manual, the instructors of the course have assembled the fundamental proteomic techniques that are taught in the 2-week course. These universal proteomic techniques can be expanded for broad use in biological research, and we hope this manual will serve as a handy reference for many scientists working at the bench.

Genomics and Proteomics

Prior to having complete genome sequences, biological research projects were primarily performed on a limited number of genes or proteins. Genetic screens or biochemical strategies were routinely used to identify new genes and proteins involved in a biological process. Isolating and sequencing genes and proteins were slow, laborious processes. Novel proteins were typically identified by Edman sequencing, which requires isolating individual proteins and identifying one amino acid at a time from peptides derived from the protein.

The advent of molecular biology, PCR, and large-scale DNA sequencing fundamentally changed biological research. Improved methods for producing recombinant proteins from cloned genes have enabled a staggering variety of functional assays and an exponential increase in the number of solved three-dimensional protein structures. Rapid and inexpensive DNA sequencing technologies have simplified the elucidation of the nucleotide sequences of large numbers of genes and genomes. Annotating this sequence information with structural and biological information has become a global venture (ENCODE Project Consortium 2004; Birney et al. 2007). Using annotated genomes, protein databases have been created that theoretically represent the complete proteome encoded by an organism's genome. This manual teaches various proteomics approaches and technologies to utilize and exploit annotated genomes in biological research.

The Challenge of the Proteome

Proteins vary greatly in size, shape, isoelectric point, hydrophobicity, and biological affinity. The diversity of amino acid side chains and the ability of proteins to fold into unique three-dimensional conformations give each protein distinct physical, chemical, and functional properties. It is this diversity that enables proteins to perform so many heterogeneous functions in the cell. In addition to the great diversity of physical properties, protein abundance in a proteome can range over six orders of magnitude. Regulatory proteins, like transcription factors, may average one to ten copies per cell, while structural proteins may be present at 1,000,000 copies per cell (Ghaemmaghami et al. 2003). In human blood, the dynamic range of protein abundance is estimated at 9–12 orders of magnitude (Omenn et al. 2005). A continuing challenge of proteomics is working with the diversity of physical properties and widely varying amounts of proteins.

Compared to its genome, an organism's proteome is considerably more complicated. Alternative RNA splicing, RNA editing, proteolytic processing, and posttranslational modifications dramatically increase the complexity of a cell's proteome (Modrek et al. 2001; Zavolan et al. 2002). It has been estimated that 40%–60% of all human genes produce transcripts with alternative exons (Modrek and Lee 2002). In a study to annotate the human genome, the ENCODE project estimates that protein-coding regions of the genome average five to six transcripts per gene (Denoeud et al. 2007). How many of these alternative transcripts are translated into proteins is unknown. Protein posttranslational modifications, like the addition of functional groups to a polypeptide, alteration of its structure, or changes to the amino acid chemistry, vastly extend the range of protein functions and activities (Walsh 2006). These modifications to proteins and protein-encoding transcripts contribute greatly to proteome diversity. For example, the ~25,000 protein-coding genes in the human genome are estimated to produce 500,000 to 2,000,000 unique proteins (International Human Genome Sequencing Consortium 2004; Jensen 2004; Birney et al. 2007).

While an organism's genome is relatively constant, its proteome is highly dynamic. An organism will have drastically different protein expression in different cell types, stages of the cell cycle, stages of development, and environmental conditions. Protein interactions, intracellular locations, and posttranslational modifications are constantly changing during all these events. One of the great challenges for proteomics is accurately and comprehensively identifying and quantifying changes in a cell or tissue's proteome.

Despite the global effort to annotate genomes and proteomes, the biological functions of most genes and proteins are unknown. For example, the yeast *Saccharomyces cerevisiae*, which has a relatively small proteome of ~6000 proteins, continues to be the focus of a number of large-scale genetic, genomic, and proteomic efforts to comprehensively annotate its genome and proteome (Winzeler et al. 1999; Uetz et al. 2000; Tong et al. 2001, 2004; Gavin et al. 2002, 2006; Giaever et al. 2002; Ho et al. 2002; Measday et al. 2005; Krogan et al. 2006). Nonetheless, we still have little knowledge of the biological function of 40% of its proteome (Clare and King 2003; Hirschman et al. 2006).

PROTEOMICS: FUNDAMENTALS, TECHNIQUES, AND APPLICATIONS

Protein Biochemistry and Purification

The foundation for proteomics is protein biochemistry. For students and scholars venturing into proteomics, a fundamental knowledge of protein biochemistry and protein purification is essential. For the users of this manual, a basic understanding of protein biochemistry and protein purification is assumed. Practical knowledge of how to harvest cells and tissues, prepare protein extracts, purify proteins, run SDS-PAGE, and detect proteins will improve your success in proteomics. For those needing more basic training, there are several excellent resources that the user can consult (Marshak et al. 1996; Walker 1996; Golemis 2002; Simpson 2003).

Displaying the Proteome

In the 1970s, two-dimensional (2D) gel electrophoresis was developed as a powerful and sensitive technique to resolve and visualize large numbers of proteins from whole-cell extracts (O'Farrell

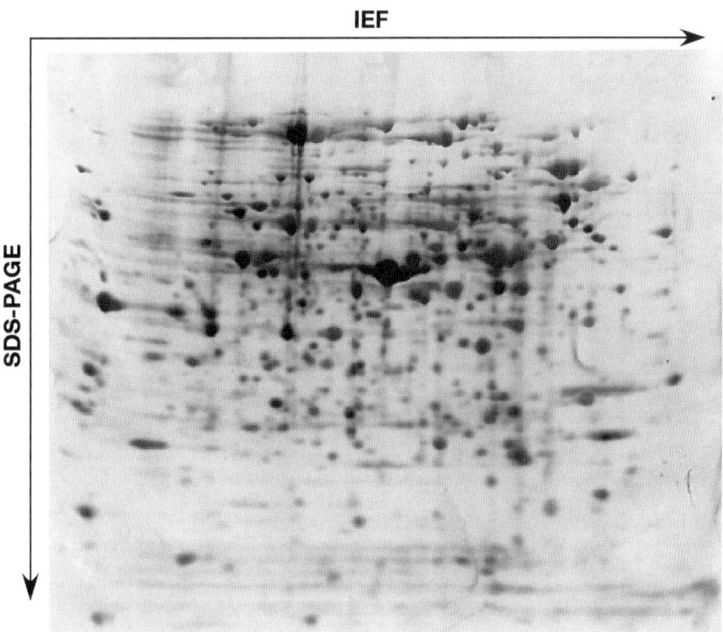

FIGURE 1. Two-dimensional (2D) gel electrophoresis display of the *Escherichia coli* proteome. The *E. coli* protein extract was prepared from cells growing in rich media. Isoelectric focusing (IEF) was first used to fractionate based on the isoelectric point of the proteins (amino acid composition). Sodium dodecyl sulfate polyacryamide gel electrophoresis (SDS-PAGE) was then used to fraction the proteins based on the molecular weight of the proteins (size). The intensity of the 2D spots reflects the abundance of the proteins in the proteome. The Coomassie Blue stained gel displays >1000 2D spots. (Reprinted from Link A.J. et al. 2004, © Elsevier 2004.)

1975; O'Farrell et al. 1977). By coupling isoelectric focusing with SDS-PAGE, thousands of proteins can be separated (Fig. 1). The intensity of 2D gel spots reflects the abundance of the proteins and is used to quantify the proteome. The reproducibility of 2D gel electrophoresis has been dramatically improved by the use of immobilized pH gradients (IPG), standardization of the procedure, fluorescent labels for proteins, and improvements in imaging and computational analysis (Bjellqvist et al. 1982; Blomberg et al. 1995; Klose and Kobalz 1995; Unlü et al. 1997; Link 1998; Wildgruber et al. 2000; Righetti et al. 2004). The technique is commonly used for globally profiling the expressed proteins from cells and tissues. To identify the proteins, the 2D spots are typically excised from the gel, in-gel digested with trypsin, and the recovered peptides are then identified by mass spectrometry. Experiment 1 trains students to set up and perform 2D gel electrophoresis analysis of whole-cell lysates, including the identification of proteins within the 2D gel spots.

Identifying and Quantifying the Proteome Using Mass Spectrometry

Mass spectrometry has been a major technology driving proteomics. It has become the method of choice for sensitively and rapidly identifying peptides, proteins, and posttranslational modifications (Fig. 2). The technique is capable of generating vast amounts of data on large numbers of proteins in a short period of time. With the addition of various approaches for labeling peptides and proteins with stable isotopes, it can also be used to quantify the abundance of large numbers of proteins in the proteome.

Formally, mass spectrometry measures the mass to charge ratio (m/z) of gas-phase ions. It is commonly used in analytical chemistry to quantify and characterize the structure of a variety of

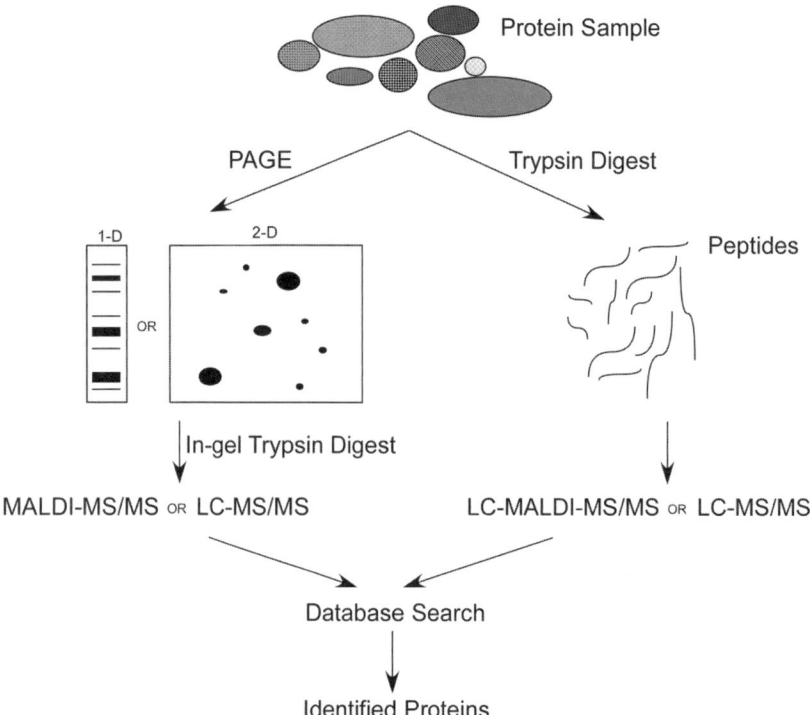

FIGURE 2. Mass spectrometry approaches to identify proteins. A protein sample (e.g., cell lysate, subcellular fraction, protein complex, or purified protein) is either fractionated by gel electrophoresis or digested directly with trypsin. For gel-fractionated proteins, the proteins are in-gel digested with trypsin and the peptides extracted. The recovered peptides are analyzed by matrix assisted laser desorption ionization (MALDI) coupled with tandem mass spectrometry (MS/MS) to acquire a sequencing spectrum of individual peptides. Alternatively, the peptides are analyzed by liquid chromatography and tandem mass spectrometry (LC-MS/MS) to acquire tandem mass spectra. For both approaches, the acquired spectral data is computationally compared to the theoretical properties of protein sequences in a database to identify the proteins in the sample.

small molecules (<1000 Da). In the late 1980s, two ionization techniques were introduced that were suitable for ionizing peptides and proteins. Matrix-assisted laser desorption ionization (MALDI) and electrospray ionization (ESI) are soft ionization techniques (Whitehouse et al. 1985; Karas and Hillenkamp 1988). They typically do not degrade polypeptides during the ionization process and are extremely efficient at generating peptide and protein ions. Using either MALDI or ESI, mass spectrometry (MS) accurately measures the m/z values of peptides and proteins. Tandem mass spectrometry (MS/MS) isolates individual peptide ions and then fragments them to generate sequencing ions, from which the m/z ratios can be used to determine the peptide's amino acid sequence and structure. Experiments 3, 4, 6, and 7, and Appendices 1 and 7 train students to perform MS/MS analysis using MALDI and ESI to identify proteins.

The coupling of liquid chromatography with MS (LC-MS) creates a powerful analytical technique for the analysis of mixtures of peptides. Liquid chromatography, especially high-performance liquid chromatography (HPLC), physically separates the peptides in a complex mixture prior to analysis by MS. The development of small-scale, low-flow-rate chromatography systems dramatically increases the sensitivity of LC-MS. Today, microcapillary liquid chromatography coupled with MS/MS dominates in the proteomic analysis of complex mixtures of peptides and proteins. It has proven to be an extremely powerful approach for rapidly identifying and quantifying peptides, proteins, and posttranslational modifications (Aebersold and Mann 2003; Cravatt et al. 2007). Experiment 4 trains students to set up and operate an ESI source for coupling reversed-phase microcapillary liquid chromatography with MS/MS to identify low-abundance trypsin-digested proteins. Experiment 7 teaches students how to couple liquid chromatography with MALDI-MS to analyze and compare proteomes.

Multidimensional liquid chromatography methods coupled with MS/MS is a powerful approach for rapidly identifying and comparing the proteomes from cells and tissues (Link et al. 1999; Washburn et al. 2001; Zhen et al. 2004). In an analogy to DNA sequencing of genomes, "shotgun proteomics" involves trypsin-digesting the proteome into peptides, fractionating the peptides using multidimensional liquid chromatography, identifying individual peptides by MS/MS, and re-assembling the identified peptide sequences into a list of proteins. By in vivo labeling of cells with isotopic amino acids or in vitro labeling of proteins extracted from cells of tissues with chemically reactive isotopic tags, quantitative comparison of proteomes can be performed using multidimensional liquid chromatography coupled with MS/MS (Gygi et al. 1999a; Ong et al. 2002; Ross et al. 2004). Experiment 6 trains students to set up and run multidimensional microcapillary liquid chromatography directly coupled with MS/MS to comprehensively analyze trypsin-digested proteomes. Experiment 7 instructs students how to label proteomes with stable isotopes and use multidimensional chromatography coupled with MALDI to quantitatively analyze and compare proteomes (Ross et al. 2004).

The processing and analysis of MS data is an essential component of proteomics. For most students, it is one of the most important and useful components in their training at the CSHL Proteomics course. Prior to the genome sequencing projects, interpretation of tandem mass spectra of peptides was performed manually. The de novo interpretation of spectra to derive the amino acid sequence of a peptide required considerable training and expertise. In 1994, the program SEQUEST was the first published program to use a protein database and uninterpreted MS/MS data to identify proteins (Eng et al. 1994). The program compares the theoretical properties of proteins in a protein database to the experimental values from MS/MS experiments. Since the publication of the Sequest algorithm, a number of other programs have been developed that use protein databases for interpreting MS data (Perkins et al. 1999; Craig and Beavis 2004; Geer et al. 2004; Tabb et al. 2007). Experiment 8 trains students to use several MS search engines to process and analyze MS/MS data from trypsin-digested proteins into a list of validated proteins and posttranslational modifications.

Quantifying the Proteome

DNA microarray profiling and other technologies enable investigators to measure the transcriptional activity and abundance of mRNAs in cells and tissues. However, the amount of mRNA expression does not measure the level of posttranscriptional regulation that ultimately determines the final protein expression levels. Several studies have revealed that mRNA levels do not necessarily correlate with protein levels (Gygi et al. 1999b; Ideker et al. 2001; Chen et al. 2002; Greenbaum et al. 2003; Washburn et al. 2003; Lu et al. 2007). Quantitative proteomics, which precisely measures the amounts of proteins in the cell, complements the information provided by measuring transcriptional activity. This manual teaches two different methods for quantifying the proteome. Experiment 1 uses 2D gel electrophoresis and Experiment 7 teaches isotopic labeling and MS techniques to quantify proteins in the proteome.

Protein Interactions

It is commonly assumed that most proteins function in complexes, not individually, to perform specific biological processes (Alberts 1998). For example, signaling pathways that transmit signals from the exterior of the cell are mediated by protein–protein interactions. The identification of protein interactions has led to the development of biological networks that attempt to map the complete set of interactions in cells (Uetz et al. 2000; Gavin et al. 2002, 2006; Ho et al. 2002; Rual et al. 2005; Krogan et al. 2006).

The interactome is a term used to describe all of the interactions in the cell (Sanchez et al. 1999; Cusick et al. 2005). Identifying protein interactions with other proteins and macromolecules (e.g., DNA sequences, RNA molecules, and small molecules) is one of the major applications of proteomic research. The functions of an uncharacterized protein can often be inferred by identifying what other molecules it interacts with it. For characterized proteins, novel functions can be recognized by the discovery of unexpected interactions with other proteins. A protein can be a component in multiple complexes, each with a unique function (Cohen 2002; Sanders et al. 2002). Experiment 2 teaches two different protocols for isolating in vivo protein complexes from cells or tissues. Experiment 11 trains students how to use protein microarrays to identify protein interactions in vitro.

Posttranslational Modifications

While the genome can predict the linear order of amino acids, the sites of posttranslational modification in proteins are largely unknown. Posttranslational modifications (PTMs) are covalent chemical modifications of proteins that typically occur after translation from mRNAs. The chemical modifications of proteins are extremely important because they potentially change a protein's physical or chemical properties, conformation, activity, cellular location, or stability. For example, a protein may be active or inactive depending on its phosphorylation state. Over 400 different protein modifications have been identified and more are likely to be identified (Creasy and Cottrell 2004). Proteomics, especially MS, is playing an increasingly dominant role in identifying protein modifications and the substrates for the modifying enzymes.

It has been estimated that 30% of the proteome is phosphorylated (Hubbard and Cohen 1993; Cohen 2001). Phosphoproteomics is a branch of proteomics that focuses on identifying, quantifying, cataloguing, and comparing the phosphoproteins in the proteome. Experiment 5 trains students in techniques that enrich for phosphopeptides from whole proteomes and in the use of MS/MS to identify the phosphorylation sites.

Enzymatic reactions are responsible for generating every posttranslational protein modification. Protein kinases, phosphatases, acetyltransferases, deacetyltransferases, methyltransferases, demethyltransferases, ubiquitin ligases, deubiquitinylating enzymes, and proteases are examples of the enzymes dedicated to posttranslational modifications. It has been estimated that 5% of the translated human genome may encode enzymes committed to PTMs (Walsh 2006). Experiments 2 and 11 teach several approaches to identify the protein substrates of modifying enzymes.

Functional Proteomics

Functional proteomics has developed to investigate the function of individual proteins. Functional proteomics typically requires the construction of comprehensive clone sets of complete open reading frames (ORFs) that represent each gene in an organism's genome. The clone collections facilitate the high-throughput, systematic expression of individual proteins. Although large-scale cDNA clones have been used to identify a large number of expressed genes along with exon–intron boundaries, the clones are not suitable for protein expression because of the presence of 5' and 3' untranslated sequences. For comprehensive protein expression, cloning the complete, full-length

ORF is required. PCR amplification of complete ORFs from genomic DNA or cDNA libraries has been used to build comprehensive sets of clones from prokaryotic and eukaryotic organisms. When recombinational cloning systems are used to clone the sequences, the ORF can be conveniently transferred to a number of different expression vectors for specific experiments (Hartley et al. 2000). Several groups have developed strategies to use the clone libraries (Braun and LaBaer 2003; Rolfs et al. 2008). Experiment 9 teaches how to perform high-throughput cloning of ORFs and assemble sets of expression constructs.

The availability of genome-scale collections of cloned ORFs has made it possible to develop high-throughput strategies for globally identifying protein interactions and biochemical activities (Martzen et al. 1999; Uetz et al. 2000). Several groups have developed strategies to use the clone libraries for building protein microarrays (Zhu et al. 2001; Ramachandran et al. 2004) (Fig. 3). Protein arrays are being assembled for a number of applications including the identification of protein–protein and protein–small molecule interactions and to profile immunoresponses (Zhu et al. 2001; Anderson et al. 2008). Experiment 10 teaches students how to construct high-density protein arrays using ORF clone sets expressed directly on glass sides. Experiment 11 teaches an approach using protein arrays to identify protein–protein interactions.

A Format for Teaching Proteomics

At the CSHL Proteomics course, we accept 16 highly qualified and motivated students for the 2-week course each year. The students are an equal mix of experienced laboratory supervisors, postdoctoral fellows, and graduate students. We divide the students into four groups, each group having a mix of experience in science and proteomics. During the 2 weeks, the four groups rotate among the different experiments described in this manual. We use a combination of general didactic instruction to the entire group and individual instruction for each group of four students. In addition, we invite eight or nine outside speakers to give lectures and to directly interact with the students. The speakers cover a broad range of proteomics topics that are either not covered in the course or reinforce the experiments that we teach. The students spend most of their time performing the experiments described in this manual under the supervision of the instructors. We strongly believe hands on experiences are the best method for teaching proteomics. This manual describes the experiments that we use to teach proteomics.

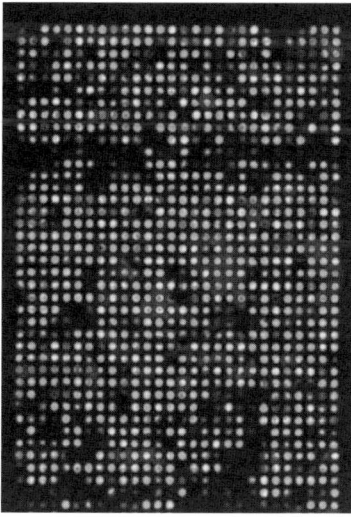

FIGURE 3. A portion of a protein microarray of individual human proteins constructed on a glass slide (see Experiment 10, Figure 13, p. 171 for a full-sized color figure). The proteins were detected using an antibody to a GST epitope fused to each protein. The protein microarray was created using the protocols in Experiments 9 and 10 of this manual.

REFERENCES

Aebersold R. and Mann M. 2003. Mass spectrometry-based proteomics. *Nature* **422:** 198–207.

Alberts B. 1998. The cell as a collection of protein machines: Preparing the next generation of molecular biologists. *Cell* **92:** 291–294.

Anderson K.S., Ramachandran N., Wong J., Raphael J.V., Hainsworth E., Demirkan G., Cramer D., Aronzon D., Hodi F.S., Harris L., et al. 2008. Application of protein microarrays for multiplexed detection of antibodies to tumor antigens in breast cancer. *J. Proteome Res.* **7:** 1490–1499.

Birney E., Stamatoyannopoulos J.A., Dutta A., Guigó R., Gingeras T.R., Margulies E.H., Weng Z., Snyder M., Dermitzakis E.T., Thurman R.E., et al. 2007. Identification and analysis of functional elements in 1% of the human genome by the ENCODE pilot project. *Nature* **447:** 799–816.

Bjellqvist B., Ek K., Righetti P.G., Gianazza E., Görg A., Westermeier R., and Postel W. 1982. Isoelectric focusing in immobilized pH gradients: Principle, methodology and some applications. *J. Biochem. Biophys. Methods* **6:** 317–339.

Blomberg A., Blomberg L., Norbeck J., Fey S.J., Larsen P.M., Larsen M., Roepstorff P., Degand H., Boutry M., Posch A., et al. 1995. Interlaboratory reproducibility of yeast protein patterns analyzed by immobilized pH gradient two-dimensional gel electrophoresis. *Electrophoresis* **16:** 1935–1945.

Braun P. and LaBaer J. 2003. High throughput protein production for functional proteomics. *Trends Biotechnol.* **21:** 383–388.

Chen G., Gharib T.G., Huang C.-C., Taylor J.M.G, Misek D.E., Kardia S.L.R., Giordano T.J., Iannettoni M.D., Orringer M.B., Hanash S.M., et al. 2002. Discordant protein and mRNA expression in lung adenocarcinomas. *Mol. Cell. Proteomics* **1:** 304–313.

Clare A. and King R.D. 2003. Predicting gene function in *Saccharomyces cerevisiae*. *Bioinformatics* (Suppl. 2) **19:** ii42–49.

Cohen P. 2001. The role of protein phosphorylation in human health and disease. The Sir Hans Krebs Medal Lecture. *Eur. J. Biochem.* **268:** 5001–5010.

Cohen P.T. 2002. Protein phosphatase 1–targeted in many directions. *J. Cell Sci.* **115:** 241–256.

Craig R. and Beavis R.C. 2004. TANDEM: Matching proteins with tandem mass spectra. *Bioinformatics* **20:** 1466–1467.

Cravatt B.F., Simon G.M., and Yates III J.R. 2007. The biological impact of mass-spectrometry-based proteomics. *Nature* **450:** 991–1000.

Creasy D.M. and Cottrell J.S. 2004. Unimod: Protein modifications for mass spectrometry. *Proteomics* **4:** 1534–1536.

Cusick M.E., Klitgord N., Vidal M., and Hill D.E. 2005. Interactome: gateway into systems biology. *Hum. Mol. Genet.* **14 Spec No. 2:** R171–181.

Denoeud F., Kapranov P., Ucla C., Frankish A., Castelo R., Drenkow J., Lagarde J., Alioto T., Manzano C., Chrast J., et al. 2007. Prominent use of distal 5′ transcription start sites and discovery of a large number of additional exons in ENCODE regions. *Genome Res.* **17:** 746–759.

ENCODE Project Consortium. 2004. ENCODE (ENCyclopedia Of DNA Elements) Project. *Science* **306:** 636–640.

Eng J.K., McCormack A.L., and Yates III J.R. 1994. An approach to correlate tandem mass spectral data of peptides with amino acid sequences. *J. Am. Soc. Mass Spectrom.* **5:** 976–989.

Gavin A.C., Aloy P., Grandi P., Krause R., Boesche M., Marzioch M., Rau C., Jensen L.J., Bastuck S., Dümpelfeld B., et al. 2006. Proteome survey reveals modularity of the yeast cell machinery. *Nature* **440:** 631–636.

Gavin A.C., Bösche M., Krause R., Grandi P., Marzioch M., Bauer A., Schultz J., Rick J.M., Michon A.M., Cruciat C.M., et al. 2002. Functional organization of the yeast proteome by systematic analysis of protein complexes. *Nature* **415:** 141–147.

Geer L.Y., Markey S.P., Kowalak J.A., Wagner L., Xu M., Maynard D.M., Yang X., Shi W., and Bryant S.H. 2004. Open mass spectrometry search algorithm. *J. Proteome Res.* **3:** 958–964.

Ghaemmaghami S., Huh W.K., Bower K., Howson R.W., Belle A., Dephoure N., O'Shea E.K., and Weissman J.S. 2003. Global analysis of protein expression in yeast. *Nature* **425:** 737–741.

Giaever G., Chu A.M., Ni L., Connelly C., Riles L., Véronneau S., Dow S., Lucau-Danila A., Anderson K., André B., et al. 2002. Functional profiling of the *Saccharomyces cerevisiae* genome. *Nature* **418**: 387–391.

Golemis E., ed. 2002. *Protein-protein interactions: A molecular cloning manual*. Cold Spring Harbor Laboratory Press, Cold Spring Harbor, New York.

Greenbaum D., Colangelo C., Williams K, and Gerstein M. 2003. Comparing protein abundance and mRNA expression levels on a genomic scale. *Genome Biol.* **4**: 117.

Gygi S.P., Rist B., Gerber S.A., Turecek F., Gelb M.H., and Aebersold R. 1999a. Quantitative analysis of complex protein mixtures using isotope-coded affinity tags. *Nat. Biotechnol.* **17**: 994–999.

Gygi S.P., Rochon, Y., Franza B.R., and Aebersold R. 1999b. Correlation between protein and mRNA abundance in yeast. *Mol Cell Biol,* **19**: 1720–1730.

Hartley J.L., Temple G.F., and Brasch M.A. 2000. DNA cloning using in vitro site-specific recombination. *Genome Res.* **10**: 1788–1795.

Hirschman J.E., Balakrishnan R., Christie K.R., Costanzo M.C., Dwight S.S., Engel S.R., Fisk D.G., Hong E.L., Livstone M.S., Nash R., et al. 2006. Genome Snapshot: A new resource at the *Saccharomyces* Genome Database (SGD) presenting an overview of the *Saccharomyces cerevisiae* genome. *Nucleic Acids Res.* **34**: D442–445.

Ho Y., Gruhler A., Heilbut A., Bader G.D., Moore L., Adams S.L., Millar A., Taylor P., Bennett K., Boutilier K., et al. 2002. Systematic identification of protein complexes in *Saccharomyces cerevisiae* by mass spectrometry. *Nature* **415**: 180–183.

Hubbard M.J. and Cohen P. 1993. On target with a new mechanism for the regulation of protein phosphorylation. *Trends Biochem. Sci.* **18**: 172–177.

Ideker T., Thorsson V., Ranish J.A., Christmas R., Buhler J., Eng J.K., Bumgarner R., Goodlett D.R., Aebersold R., and Hood L. 2001. Integrated genomic and proteomic analyses of a systematically perturbed metabolic network. *Science* **292**: 929–934.

International Human Genome Sequencing Consortium. 2004. Finishing the euchromatic sequence of the human genome. *Nature* **431**: 931–945.

Jensen O.N. 2004. Modification-specific proteomics: Characterization of post-translational modifications by mass spectrometry. *Curr. Opin. Chem. Biol.* **8**: 33–41.

Karas M. and Hillenkamp F. 1988. Laser desorption ionization of proteins with molecular masses exceeding 10,000 daltons. *Anal. Chem.* **60**: 2299–2301.

Klose J. and Kobalz U. 1995. Two-dimensional electrophoresis of proteins: an updated protocol and implications for a functional analysis of the genome. *Electrophoresis* **16**: 1034–1059.

Krogan N.J., Cagney G., Yu H., Zhong G., Guo X., Ignatchenko A., Li J., Pu S., Datta N., Tikuisis A.P., et al. 2006. Global landscape of protein complexes in the yeast *Saccharomyces cerevisiae*. *Nature* **440**: 637–643.

Link A.J., ed. 1998. *2-D Protocols for proteome analysis*. Humana Press, Inc., Totowa, New Jersey.

Link A.J. 2004. Complex mixture analysis. In *Encyclopedia of mass spectrometry 2*. (ed. M.I. Gross and R.M. Caprioli), chapter 3, pp. 274–289. Elsevier Science, New York.

Link A. J., Eng J., Schieltz D.M., Carmack E., Mize G.J., Morris D.R., Garvik B.M., and Yates III J.R. 1999. Direct analysis of protein complexes using mass spectrometry. *Nat. Biotechnol.* **17**: 676–682.

Lu P., Vogel C., Wang R., Yao X., and Marcotte E.M. 2007. Absolute protein expression profiling estimates the relative contributions of transcriptional and translational regulation. *Nat. Biotechnol.* **25**: 117–124.

Marshak D.R., Kandonaga J.T., Burgess R.R., Knuth M.W., Brennan W.A. Jr., and Lin S.-H. 1996. *Strategies for protein purification and characterization: A laboratory course manual*. Cold Spring Harbor Laboratory Press, Cold Spring Harbor, New York.

Martzen M.R., McCraith S.M., Spinelli S.L., Torres F.M., Fields S., Grayhack E.J., and Phizicky E.M. 1999. A biochemical genomics approach for identifying genes by the activity of their products. *Science* **286**: 1153–1155.

Measday V., Baetz K., Guzzo J., Yuen K., Kwok T., Sheikh B., Ding H., Ueta R., Hoac T., Cheng B., et al. 2005. Systematic yeast synthetic lethal and synthetic dosage lethal screens identify genes required for chromosome segregation. *Proc. Natl. Acad. Sci.* **102**: 13956–13961.

Modrek B. and Lee C. 2002. A genomic view of alternative splicing. *Nat. Genet.* **30**: 13–19.

Modrek B., Resch A., Grasso C., and Lee C. 2001. Genome-wide detection of alternative splicing in expressed sequences of human genes. *Nucleic Acids Res.* **29:** 2850–2859.

O'Farrell P.H. 1975. High resolution two-dimensional electrophoresis of proteins. *J. Biol. Chem.* **250:** 4007–4021.

O'Farrell P.Z., Goodman H.M., and O'Farrell P.H. 1977. High resolution two-dimensional electrophoresis of basic as well as acidic proteins. *Cell* **12:** 1133–1141.

Omenn G.S. 2006. Strategies for plasma proteomic profiling of cancers. *Proteomics* **6:** 5662–5673.

Omenn G.S., States D.J., Adamski M., Blackwell T.W., Menon R., Hermjakob H., Apweiler R., Haab B.B., Simpson R.J., Eddes J.S., et al. 2005. Overview of the HUPO Plasma Proteome Project: results from the pilot phase with 35 collaborating laboratories and multiple analytical groups, generating a core dataset of 3020 proteins and a publicly-available database. *Proteomics* **5:** 3226–3245.

Ong S.E., Blagoev B., Kratchmarova I., Kristensen D.B., Steen H., Pandey A., and Mann M. 2002. Stable isotope labeling by amino acids in cell culture, SILAC, as a simple and accurate approach to expression proteomics. *Mol. Cell. Proteomics* **1:** 376–386.

Perkins D.N., Pappin D.J.C., Creasy D.M., and Cottrell J.S. 1999. Probability-based protein identification by searching sequence databases using mass spectrometry data. *Electrophoresis* **20:** 3551–3567.

Ramachandran N., Hainsworth E., Bhullar B., Eisenstein S., Rosen B., Lau A.Y., Walter J.C., and LaBaer J. 2004. Self-assembling protein microarrays. *Science* **305:** 86–90.

Righetti P.G, Castagna A., Antonucci F., Piubelli C., Cecconi D., Campostrini N., Antonioli P., Astner H., and Hamdan M. 2004. Critical survey of quantitative proteomics in two-dimensional electrophoretic approaches. *J. Chromatogr. A.* **1051:** 3–17.

Rolfs A., Hu Y., Ebert L., Hoffmann D., Zuo D., Ramachandran N., Raphael J., Kelley F., McCarron S., Jepson D.A., et al. 2008. A biomedically enriched collection of 7000 human ORF clones. *PLoS ONE* **3:** e1528.

Ross P.L., Huang Y.N., Marchese J.N., Williamson B., Parker K., Hattan S., Khainovski N., Pillai S., Dey S., Daniels S., et al. 2004. Multiplexed protein quantitation in *Saccharomyces cerevisiae* using amine-reactive isobaric tagging reagents. *Mol. Cell. Proteomics* **3:** 1154–1169.

Rual J.-F., Venkatesan K., Hao T., Hirozane-Kishikawa T., Dricot A., Li N., Berriz G.F., Gibbons F.D., Dreze M., Ayivi-Guedehoussou N., et al. 2005. Towards a proteome-scale map of the human protein–protein interaction network. *Nature* **437:** 1173–1178.

Sanchez C., Lachaize C., Janody F., Bellon B., Roder L., Euzenat J., Rechenmann F., and Jacq B. 1999. Grasping at molecular interactions and genetic networks in Drosophila melanogaster using FlyNets, an Internet database. *Nucleic Acids Res.* **27:** 89–94.

Sanders S.L., Jennings J., Canutescu A., Link A.J., and Weil P.A. 2002. Proteomics of the eukaryotic transcription machinery: Identification of proteins associated with components of yeast TFIID by multidimensional mass spectrometry. *Mol. Cell. Biol.* **22:** 4723–4738.

Simpson R.J. 2003. *Proteins and proteomics: A laboratory manual.* Cold Spring Harbor Laboratory Press, Cold Spring Harbor, New York.

Tabb D.L., Fernando C.G., and Chambers M.C. 2007. MyriMatch: Highly accurate tandem mass spectral peptide identification by multivariate hypergeometric analysis. *J. Proteome Res.* **6:** 654–661.

Tong A.H., Evangelista M., Parsons A.B., Xu H., Bader G.D., Pagé N., Robinson M., Raghibizadeh S., Hogue C.W., Bussey H., et al. 2001. Systematic genetic analysis with ordered arrays of yeast deletion mutants. *Science* **294:** 2364–2368.

Tong A.H., Lesage G., Bader G.D., Ding H., Xu H., Xin X., Young J., Berriz G.F., Brost R.L., Chang M., et al. 2004. Global mapping of the yeast genetic interaction network. *Science* **303:** 808–813.

Uetz P., Giot L., Cagney G., Mansfield T.A., Judson R.S., Knight J.R., Lockshon D., Narayan V., Srinivasan M., Pochart P., et al. 2000. A comprehensive analysis of protein–protein interactions in *Saccharomyces cerevisiae*. *Nature* **403:** 623–627.

Unlü M., Morgan M.E., and Minden J.S. 1997. Difference gel electrophoresis: A single gel method for detecting changes in protein extracts. *Electrophoresis* **18:** 2071–2077.

Walker J.M., ed. 1996. *The protein protocols handbook.* Humana Press, Totowa, New Jersey.

Walsh C.T. 2006. *Posttranslational modification of proteins: Expanding nature's inventory.* Roberts and Company, Greenwood Village, Colorado.

Washburn M.P., Wolters D., and Yates III J.R. 2001. Large-scale analysis of the yeast proteome by multidimensional protein identification technology. *Nat. Biotechnol.* **19**: 242–247.

Washburn M.P., Koller A., Oshiro G., Ulaszek R.R., Plouffe D., Deciu C., Winzeler E., and Yates III J.R. 2003. Protein pathway and complex clustering of correlated mRNA and protein expression analyses in *Saccharomyces cerevisiae*. *Proc. Natl. Acad. Sci.* **100**: 3107–3112.

Whitehouse C.M., Dreyer R.N., Yamashita M., and Fenn J.B. 1985. Electrospray interface for liquid chromatographs and mass spectrometers. *Anal. Chem.* **57**: 675–679.

Wildgruber R., Harder A., Obermaier C., Boguth G., Weiss W., Fey S.J., Larsen P.M., and Görg A. 2000. Towards higher resolution: Two-dimensional electrophoresis of *Saccharomyces cerevisiae* proteins using overlapping narrow immobilized pH gradients. *Electrophoresis* **21**: 2610–2616.

Winzeler E.A., Shoemaker D.D., Astromoff A., Liang H., Anderson K., Andre B., Bangham R., Benito R., Boeke J.D., Bussey H., et al. 1999. Functional characterization of the *S. cerevisiae* genome by gene deletion and parallel analysis. *Science* **285**: 901–906.

Zavolan M., van Nimwegen E., and Gaasterland T. 2002. Splice variation in mouse full-length cDNAs identified by mapping to the mouse genome. *Genome Res.* **12**: 1377–1385.

Zhen Y., Xu N., Richardson B., Becklin R., Savage J.R., Blake K., and Peltier J.M. 2004. Development of an LC-MALDI method for the analysis of protein complexes. *J. Am. Soc. Mass Spectrom.* **15**: 803–822.

Zhu H., Bilgin M., Bangham R., Hall D., Casamayor A., Bertone P., Lan N., Jansen R., Bidlingmaier S., Houfek T., et al. 2001. Global analysis of protein activities using proteome chips. *Science* **293**: 2101–2105.

EXPERIMENT 1

Analysis of Whole-cell Lysates by Two-dimensional Gel Electrophoresis and MALDI Mass Spectrometry

Sarah L. Volk

*Department of Biological Chemistry, University of Michigan,
Ann Arbor, Michigan 48109*

Two-dimensional gel electrophoresis has long been the method of choice for the resolution of proteins from complex mixtures. It combines two well-behaved, orthogonal separation methods (isoelectric focusing and SDS gel electrophoresis) to provide the highest resolution method for separating proteins. Two-dimensional electrophoresis also has the virtues of being a stable methodology that is applicable to a large number of sample types, can be used quantitatively, has a reasonable loading capacity, provides parallel separation of proteins, and can easily be run in most research laboratories.

Two-dimensional (2D) electrophoresis is a valuable tool for proteomics research where complex protein mixtures can be separated for analysis. Two-dimensional electrophoresis facilitates the identification of proteins that have been conditionally expressed, their levels of expression, and determination of whether posttranslational modifications are affected. With the use of image analysis software packages, two or more different states of a cell or an organism can be compared allowing the qualitative and quantitative examination of proteomes. The basic experimental protocol typically followed in a two-dimensional electrophoresis experiment is the following:

1. Sample preparation
2. First dimension: isoelectric focusing (IEF)
3. Second dimension: SDS-PAGE
4. Staining
5. Imaging
6. Image analysis
7. Tryptic digestion
8. Protein identification by mass spectrometry

Sample Preparation

Good sample preparation, though often overlooked, is essential for good quality 2D results. The optimum sample preparation procedure must be experimentally determined for each sample type. Solubilization of the sample should prevent artifact formation, prevent aggregation, and promote denaturation and reduction of all the proteins in the sample in the minimum number of steps pos-

TABLE 1. Cell disruption methods

Disruption method	Application	Severity of lysis
Osmotic lysis	Blood cells, tissue culture cells	gentle
Freeze-thaw lysis	Bacterial cells, tissue culture cells	gentle
Detergent lysis	Tissue culture cells	gentle
Enzymatic lysis	Plant tissue, bacterial cells, fungal cells	gentle
Sonication	Cell suspensions	vigorous
French pressure cell	Bacteria, algae, yeasts	vigorous
Grinding	Solid tissues, microorganisms	vigorous
Mechanical homogenization	Solid tissues	vigorous
Glass bead homogenization	Cell suspensions, organisms with cell walls	vigorous

sible. The effectiveness of solubilization depends on the choice of cell disruption method, protein concentration and dissolution method, choice of detergents, and composition of the sample solution. To fully analyze all intracellular proteins, cells must be effectively disrupted. The choice of disruption method depends on the source of the sample. Table 1 summarizes some of the types and applications of lysis methods.

The presence of DNA and RNA in a sample can result in poor focusing in the acidic region of the IEF gel and can also generate a significant amount of background in a silver stained gel. Digesting the sample with a DNase/RNase cocktail or Benzonase is simple and effective. Note that DNase requires magnesium and neither DNase nor RNase are effective in urea. Benzonase will digest both DNA and RNA and retains enough activity in 8 M urea to remain effective. Other methods for DNA/RNA removal include the use of streptomycin and polyethyleneimine.

Precipitation of the sample is an optional but useful step for the removal of interfering substances and for concentration of the sample. Salts, buffers, small ionic molecules, ionic detergents, nucleic acids, polysaccharides, lipids, and phenolic compounds can interfere with the electrophoretic separation of proteins if they are present in high enough concentrations. Commercially available sample clean-up/precipitation kits are available from a number of manufacturers, including GE Healthcare, Sigma-Aldrich, and Bio-Rad.

Composition of the Solubilization Solution

To ensure complete solubilization and denaturation of proteins, the solution must contain certain components including:

- **Urea,** a neutral chaotrope, to solubilize and unfold proteins to their fully random conformation. The addition of thiourea improves solubilization, particularly of membrane proteins.
- **Nonionic or zwitterionic detergents,** such as CHAPS, ASB-14, Triton X-100, and NP-40, to ensure complete sample solubilization and prevent aggregation.
- **Reducing agents,** such as TCEP or TBP, to reduce disulfide bonds. TBP and TCEP are nonthiol reducing agents thus avoiding the pK of the thiolate (pK>9) and the necessity to account for the additional thiols during alkylation.
- **Carrier ampholytes** are used to enhance protein solubility by minimizing protein interaction due to charge–charge interactions.
- **Bromophenol Blue** allows the visual tracking of the flow of ions towards the anode.

Proteins precipitated by acid or most organic solvents are denatured and are often difficult to completely resolubilize. This is not usually a major issue for 2D gels due to the use of very efficient solubilizing methods.

> **General points to remember when preparing samples**
>
> - Keep sample preparation as simple as possible to avoid protein losses.
> - Cells should be disrupted in a way that minimizes proteolysis.
> - Solubilization solution should be made fresh or stored frozen.
> - Prepare samples just prior to IEF to preserve sample quality.
> - Prior to IEF, remove all particulate matter by ultracentrifugation.
> - To avoid modification of proteins, do not heat samples above 37°C when the solubilization solution contains urea.

First Dimension Basic Principles

Isoelectric focusing (IEF) is electrophoresis in a pH gradient that increases in pH from anode to cathode. In contrast to conventional electrophoresis, the only species that can be separated by IEF are amphoteric molecules such as proteins and peptides, which contain both carboxylic and amino groups. A protein exhibits a net positive or negative charge that is dependant on the pH of its environment. When subjected to an electrical field in a pH gradient, a protein will migrate according to its charge, as in conventional electrophoresis. However, as it migrates through the pH gradient, the protein will eventually reach a pH where it has a net charge of zero, i.e., the isoelectric point (pI). Diffusion of the protein from its pI results in acquisition of charge and subsequent migration in the electric field back to its pI. This focusing effect is what makes IEF the highest resolution single method for resolving proteins. IEF provides the ability to separate components that differ in pI by only 0.001 of a pH unit by using narrow pH range gels. The proteins in the mixture are thus separated and concentrated, or focused, at their respective pI value. The pI of a protein is determined by the population of charged side chains of the amino acids. In the denaturing environment typically used for 2D PAGE in proteomic analysis, proteins are unfolded and all ionizable side chains can fully participate in determination of the pI. In their "native" configuration, proteins are folded with ionizable side chains primarily, but not exclusively, on the surface of the molecule and the pI can differ somewhat from that of the unfolded, denatured protein.

IEF is the only electrophoretic method where all the constituents reach an equilibrium position, which means that the final result is not affected by the initial distribution of the sample components. This property allows sample to be incorporated into the gel matrix at any position along its length. Commercially available immobilized pH gradient (IPG) gels are supplied in a dry state and need to be rehydrated. Protein samples are typically introduced into the IPG strips in one of two ways: protein may be included in the rehydration/solubilization solution or it may be applied directly to the rehydrated IPG strip using sample cups or sample wells. The benefits of sample loading by in-gel rehydration include the ability to load greater quantities of protein or to load dilute solutions of sample (see Table 2).

TABLE 2. Approximate protein loads and volumes for IPG strips

IPG strip length (cm)	Analytical loading (silver or SYPRO Ruby staining)	Preparative loading (Coomassie staining)	Rehydration solution volume (μL)
7	10–100 μg	100–300 μg	160
11	20–200 μg	100–500 μg	200
17/18	100–1000 μg	250 μg–1 mg	350

Equilibration

Focused IPG strips must be equilibrated in SDS-PAGE buffer prior to loading on the second dimension gel. The proteins must be coated with SDS to ensure proper migration in the second dimension. Some proteins precipitate at their pI and can appear as white bands on the focused IPG strip. The addition of urea to the SDS-PAGE equilibration buffer can resolubilize most proteins. Typically, the IPG strips are equilibrated in two steps. The first step is incubation in SDS-PAGE-urea buffer with a reducing reagent (usually TCEP or TBP) to cleave intra and interchain disulfide bonds. Iodoacetamide is added in place of the reducing reagent in the second stage of equilibration to alkylate all of the cysteine residues.

If using DTT as the reducing reagent in the equilibration step, sufficient iodoacetamide must be used to alkylate the thiols present in DTT. Insufficient alkylation of thiols can cause streaking in the second dimension. Equilibration with nonthiol reducing agents such as tributylphosphine (TBP) or TCEP can eliminate this problem.

Second Dimension Basic Principles

The second dimension SDS-PAGE step is an electrophoretic method for the separation of proteins according to their size. The proteins resolved in the IPG strip are applied to the second dimension gel and separated by size perpendicular to the first dimension. The proteins are in effect sieved, according to size, by the pores of the second dimension gel. This is made possible by SDS, an ionic detergent, which binds to the protein forming an anionic complex with relatively constant net negative charge per unit mass. The mobility of the proteins is related logarithmically to the size. Each spot in the resulting gel usually corresponds to a single protein. However, the identification of two or more different proteins within one spot is not uncommon (~10% on large format gels). Up to 10,000 individual protein species have been resolved in a single gel using this technique and information such as protein pI and molecular weight can be determined.

Visualization of Proteins

In order to visualize the separated proteins in a gel, they must be stained. This is usually done with dye or metal-based methods. When choosing a staining method, several factors must be considered: sensitivity of the stain, linear range, ease of use, expense, and compatibility with mass spectrometry. All stains interact differently with different proteins. No stain will universally stain all proteins in proportion to their amount in a gel. However, ruthenium-based fluorescence stains and colloidal Coomassie Blue are able to stain a broad spectrum of proteins and are compatible with mass spectrometry. Some characteristics of various protein stains are listed in Table 3.

TABLE 3. Sensitivity of protein stains

Staining Method	Sensitivity	Compatibility
Flamingo	0.5 ng	MS compatible, broad linear range, high sensitivity
SYPRO Ruby	1 ng	MS compatible, linear over 3 orders of magnitude, high sensitivity
Coomassie G-250	<10 ng	MS compatible, easily visualized
Silver Stain	1 ng	Vorum silver stain method is MS compatible, high sensitivity

Imaging and Image Analysis

In order to perform image analysis of a 2D gel, the image is digitized using a densitometer, CCD camera, phosphorimager, or fluorescence scanner. The goal is to achieve grayscale images that are not manually altered or saturated. Imaged gels are compared by electronic superimposition of gel images from each sample. Gels are superimposed using specialized image analysis software that allows landmark proteins to be selected and used to warp images to match. It is important to run replicate gels (three or more of each sample condition) in order to obtain accurate reproducibility and quantitation.

Image analysis software packages are necessary for the evaluation of complex 2D gels yielding comprehensive qualitative and quantitative data. The majority of image analysis software allows:

- Storage of large amounts of experimental image data
- Automatic spot detection and quantitation
- Composition of master gels from replicate 2D gels
- Comparison of 2D gel images

Protein Identification

2D gels allow protein quantitation, parallel protein purification, and are an effective starting point for protein identification by mass spectrometry. The effectiveness of 2D gels is enhanced by new instruments with greater sensitivity and throughput. The identification process initially involves image analysis and determination of the protein spots of interest, i.e., those protein spots whose changes in level are significant. These spots are excised from the gel either manually, using a dermal punch, or by an automated spot cutter. Proteins in the excised gel spots are enzymatically digested, typically with trypsin, to release peptides from the gel. Manual protein digestion is a time consuming process and can lead to contamination of samples, particularly by keratin. Automation of this process allows for high-throughput analysis and minimization of sample contamination. The resultant peptide mixture is extracted from the gel spots and analyzed using MALDI/MS to generate a peptide mass fingerprint. These peptide mass fingerprints can then be used to search against the appropriate species protein database to identify the protein(s) having the closest fit to the data. When a peptide mass fingerprint leads to an ambiguous protein identity, protein databases must be searched with a higher level of information. This information is typically obtained by tandem MS analysis of at least two peptides to determine the sequence information. Mass spectrometers that are able to generate such information include ESI LC/MS/MS and MALDI/MS/MS instruments.

Experimental Overview

The eukaryote *Saccharomyces cerevisiae* is able to switch from fermentation to respiration with major changes in metabolic activity. This switch is called the diauxic shift. Glucose in abundance is catabolized primarily by fermentation. A transient growth arrest occurs as the glucose becomes exhausted. During the growth arrest, the metabolism changes from fermentation to respiration. Alternative carbon sources are then used, such as ethanol, glycerol, and oleic acid.

In this experiment, the 2D gel profiles of yeast cells grown in the presence of glucose (fermentation) will be compared to profiles of yeast cells grown in the presence of glycerol (respiration). In the course, *S. cerevisiae* strain S288C is cultivated on rich YPD. The respiration sample is

> **Preventing sample contamination**
>
> Avoid contamination of your samples at all times by heeding the following suggestions.
>
> - Always wear powder-free gloves when handling samples, reagents, IPG strips, SDS polyacrylamide gels, and any equipment associated with the experiment.
> - Maintain a clean work environment. Clean all equipment with a suitable detergent and rinse with distilled H_2O. Work in a hood when possible.
> - Always use electrophoresis grade or the highest purity reagents and ultra pure H_2O (when required).
> - Do not wear wool or silk clothing.

the strain S288C grown on YPD and then transferred to YPG. Cells are harvested after 16 hours. After cell lysis and protein solubilization, protein samples are prepared for 2D gel electrophoresis. The 2D gels are 11-cm gels with a pH range from 4 to 7. Spot profiles of the control (i.e., fermentation) versus the experimental (i.e., respiration) are compared and differentially expressed proteins are analyzed.

PROTOCOL 1

Preparation of Yeast Cell Lysates

MATERIALS

CAUTION: See Appendix 11 for appropriate handling of materials marked with <!>.

Reagents

2D Cleanup Kit (GE Healthcare 80-6484-51)
2D Quant Kit (GE Healthcare 80-6483-56)
Ampholytes, pH 4–7
DNase/RNase solution
    1 mg/mL DNase I
    0.25 mg/mL RNase
    0.5 M Tris-Cl (pH 7.0)
    50 mM $MgCl_2$ <!>
Lysis solution
    Prepare the lysis solution by combining 8 mL CelLytic-Y (Sigma-Aldrich C4482), 200 µL 200 mM Tris-carboxyethyl phosphine <!> (TCEP stored in 1.5 M Tris-Cl [pH 8.8]), 100 µL DNase/RNase solution, and two protease inhibitor tablets (nonEDTA; Roche 10946900).
Sample solubilization/rehydration buffer
    Destreak (GE Healthcare 17-6003-18), a proprietary formulation
        A homemade alternative to Destreak is the following sample solubulization/rehydration buffer:
            7 M urea <!>
            2 M thiourea <!>
            1% ASB-14
            1% Triton X-100 <!>
            1% CHAPS <!>

ANALYSIS OF WHOLE-CELL LYSATES | 19

Prepare in deionized H$_2$O. Add a very small amount of bromophenol blue and adjust with H$_2$O to the desired final volume. Just before use, add carrier ampholytes and a reducing agent as per Step 7.

200 mM TCEP <!>

Yeast cells

In the course, *S. cerevisiae* strain S288C is cultivated on rich YPD (1% w/v yeast extract, 2% w/v peptone, and 2% w/v glucose) and grown to an optical density (600 nm) of 1.1. The respiration sample is the strain S288C grown on YPD to an optical density (600 nm) of 1–2, cells are pelleted, and transferred to YPG (3% v/v glycerol instead of glucose). Cells are harvested after 16 hours and frozen.

Equipment

Incubator set at 30°C

PROCEDURE

Cell Lysis

1. If necessary, thaw the cell pellets.
2. Add 4 mL of cold lysis solution to each thawed cell pellet.
3. Incubate for 30 minutes with gentle shaking at room temperature.
4. Centrifuge the lysed cells for 10 minutes at 4°C and 12,000g to pellet cellular debris.
5. Estimate the protein concentration using the 2D Quant Kit.

   In the course, we typically estimate ~10 μg/μL of cells. When preparing large amounts of protein, transfer the protein-containing supernatant to fresh tubes and store in aliquots of 50 μL, 200 μL, and 500 μL.

6. Precipitate the proteins using the 2D Cleanup Kit according to the manufacturer's protocol.

Protein Solubilization

7. To each 1 mL of solubilization buffer, add 10 μL of 200 mM TCEP and 10 μL of the pH 4–7 ampholytes.

   Final concentrations: 2 mM TCEP, 1% ampholytes.

8. Suspend the protein pellet in the appropriate volume of solubilization solution containing the added TCEP and ampholytes.

   A sample tube containing ~225 μg of protein requires 750 μL of solubilization solution. In the course, three gels are run from each tube, therefore, ~75 μg of protein is loaded per gel. Some losses will have occurred as a result of the precipitation step.

9. Vortex vigorously to solubilize the protein pellet.
10. Incubate for 1 hour at 30°C to aid in solubilization of the pellet. Vortex occasionally during the 1-hour incubation.
11. After incubation, centrifuge at 10,000 rpm for 5 minutes to pellet any insoluble material. The supernatant is the protein sample to be used in Protocol 2.

EXPERIMENT 1

PROTOCOL 2

IPG Strip Rehydration and Isoelectric Focusing (IEF)

MATERIALS

Reagents

Mineral oil
Protein sample (supernatant from Step 11, Protocol 1)

Equipment

IPG strips
Focusing unit (Bio-Rad)
Forceps
Protean IEF cell (Bio-Rad)
Rehydration/focusing trays
Wicks for IEF

PROCEDURE

Active Rehydration of IPG Strips

1. Orient the rehydration/focusing tray with the + sign on the left, and apply 200 µL of the yeast protein sample solution (from Step 11, Protocol 1) to a channel. Avoid forming large air bubbles (Fig. 1).

2. Grasp an IPG strip and gently fold back the protective cover. Use forceps to grasp the basic end of the plastic-backed IPG strip and peel off the protective cover.

    The plastic backing of the gel is labeled with the pH range and the acidic (anodic) end is indicated with a (+).

3. Starting with the acidic end (+) at the left, lower the IPG strip gently, **gel side down,** into the channel with solution ensuring that no bubbles are trapped under the gel. Alternately, the strip can be applied to the sample in a "U" shape from the center out to the edges. Gently move the strip to make adjustments of sample distribution or to remove bubbles (Fig. 2).

4. Pipette a layer of mineral oil on top of each IPG strip (~1 mL of mineral oil). Place the plastic lid on the focusing tray. Using a felt-tipped marker, indicate on the plastic lid which samples are located in which channels. Place the focusing tray into the focusing unit, making sure that the electrodes on the tray are making contact with the electrodes on the unit.

5. Turn on the Protean IEF cell and use the following protocol to perform active rehydration:

    Select "Rehydration"

    Rehydration Condition: "Active @ 50 V"

    Temperature: "25°C"

    "Start"

    Rehydrate for a minimum of 16 hours (Fig. 3).

FIGURE 1. The yeast protein solution is pipetted into a channel of the rehydration tray.

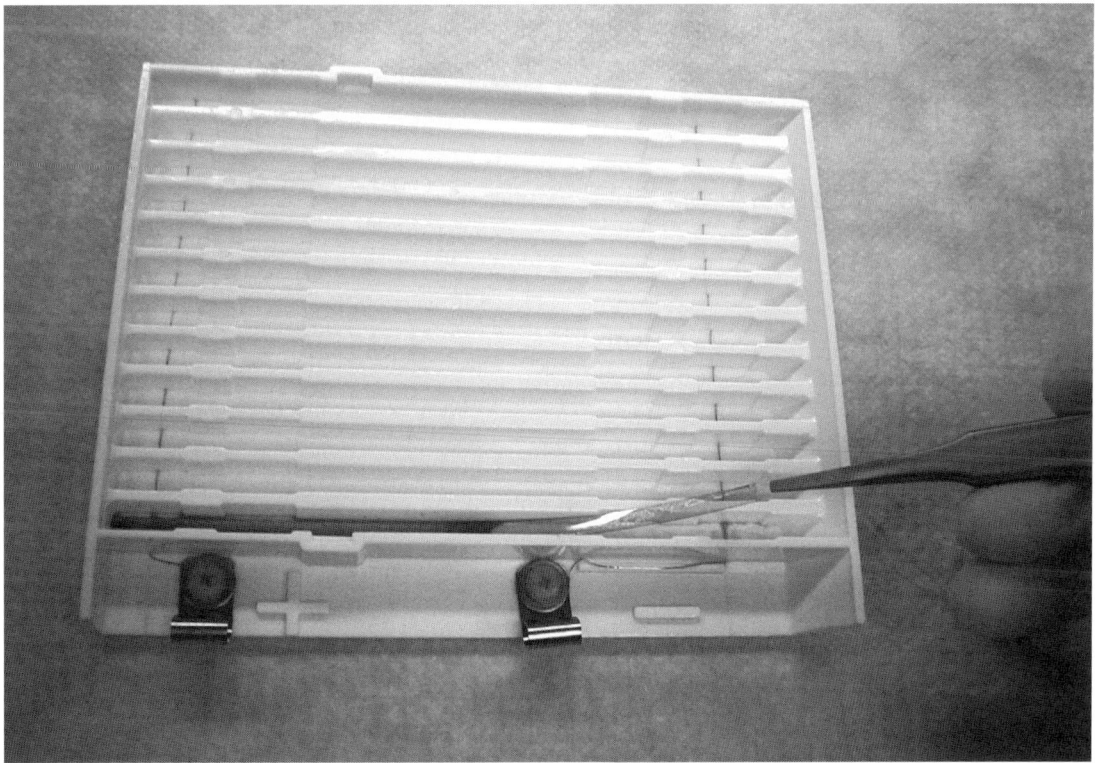

FIGURE 2. Using forceps, the immobilized pH gradient (IPG) strip is gently lowered over the sample solution. The gel is facing downward, and care is taken to avoid forming air bubbles underneath the strip.

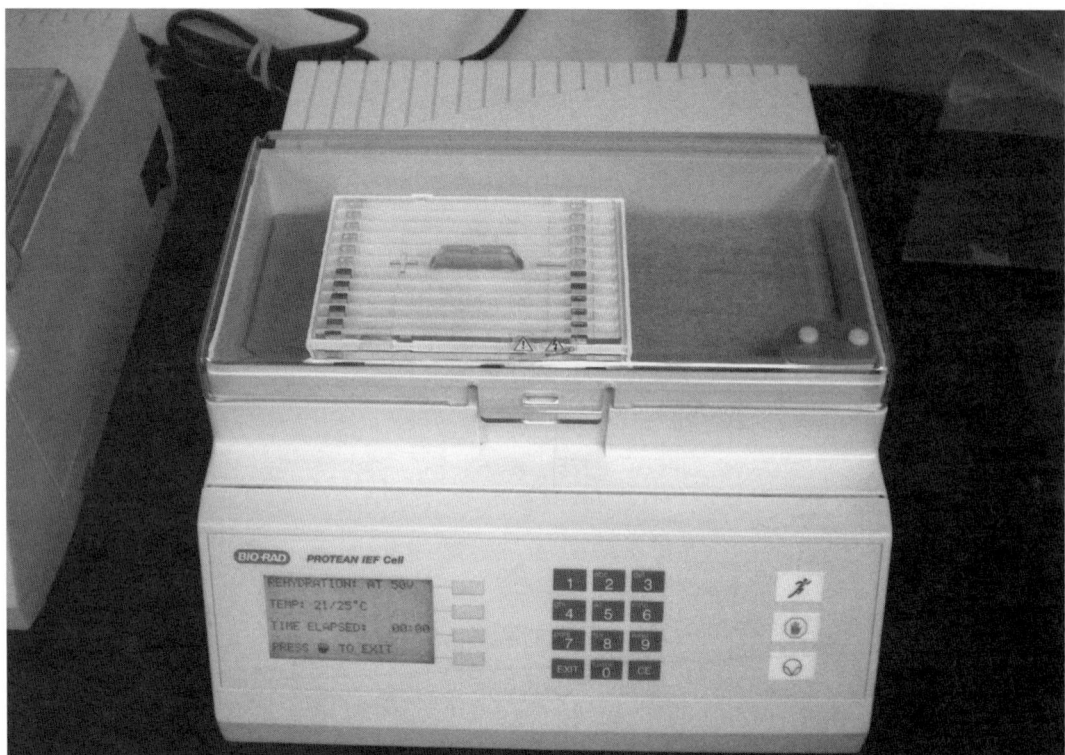

FIGURE 3. Active rehydration takes places for 16–24 hours at a constant voltage of 50 V.

First Dimension IEF

6. Stop active rehydration by pressing "Stop" when the timer has indicated that the rehydration has reached at least 16 hours. Remove the focusing tray from the Protean IEF cell.

7. Using forceps, place the desired number of wicks, two per strip, onto a Kimwipe. Pipette 40–50 μL of H_2O onto each wick. Blot off the excess H_2O.

8. Using forceps, gently lift up the end of the rehydrated IPG strip by the plastic backing, and use forceps to place a wick on each metal electrode in each channel of the focusing tray. Take care not to touch the rehydrated gel (Fig. 4).

   The electrode wicks serve to absorb materials that are eluted from the strips during focusing and should be changed frequently. Changing wicks frequently will reduce the resistance, improve the voltage ramp rate, and help reduce the buildup of denatured proteins at the end of the strip. Make sure that the strip is positioned in the channel so that the gel portion of the strip makes contact directly over an electrode and is on top of the dampened wick.

9. Replace the cover onto the tray. Place the focusing tray in the Protean IEF cell. Again make sure that the electrodes on the tray make proper contact with the electrodes in the unit.

10. Program the IEF cell using the conditions shown below. Use the default cell temperature of 20°C. **The maximum current limit per strip is 50 μA.** The run time is determined from accumulated volt-hours.

    Focusing conditions for yeast samples:

    250 V for 15 minutes

    Voltage ramp up to 8000 V over the course of 2 hours 30 minutes

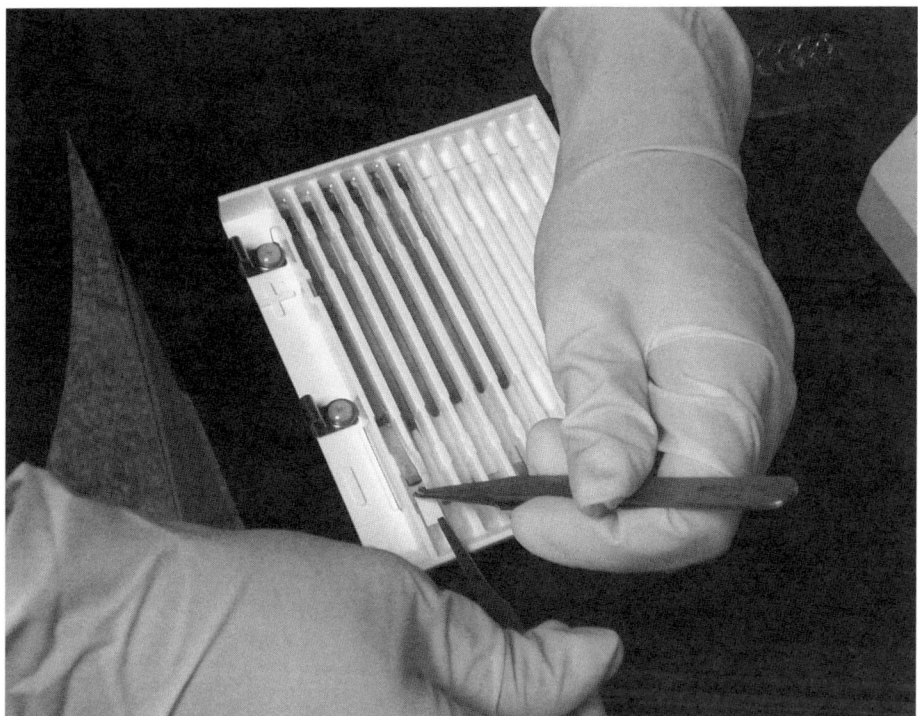

FIGURE 4. During the isoelectric focusing, the program is paused, the IPG strip is lifted one end at a time, and a dampened electrode wick is placed on top of the electrode wire. The IPG gel then sits directly on top of the electrode wick.

50,000 volt-hours

50 µA per strip

20°C

The Bio-Rad Protean IEF cell comes with preprogrammed methods. If using a manual program, use Table 4 as a guideline for an 11-cm IPG strip. Remember that different samples may require more or less volt hour accumulation.

11. Change the wicks 60–120 minutes after beginning the run. Hit the "Pause" (blue) button—*NOT the red Stop button!*—on the unit and wait until the voltage drops to zero. Lift the lid and remove the focusing tray from the unit. Remove the cover and, using forceps, lift the ends of each strip and remove each wick. Replace with fresh, dampened wicks in the same fashion. Replace the cover and hit the Run button to resume. Repeat later in the run as needed.

At the beginning of the run, the resistance in each strip is low and the current will be held at the maximum, 50 µA/strip, hence the programmed voltage may not be reached. As debris is eluted, the current will typically begin to drop and the voltage will increase; the low value for

TABLE 4. Voltage Steps for IEF on 11-cm Strips

Volts	Time (minutes)	Volt-hours
250	15	62.5
500	15	125
1000	15	250
2000	30	1000
4000	30	2000
8000	372	49,600
	Total Volt-hours	53,038

TABLE 5. Maximum voltages for different strip lengths

IPG strip length (cm)	Maximum voltage (V)	Approximate volt-hours	Maximum current per strip (µA)
7	3000	2500–20,000	50
11	8000	35,000–80,000	50
17 or 18	10,000	60,000–120,000	50

each strip is typically 10 µA/strip. Samples with excessive amounts of salts and proteins outside of the focusing range generally require more frequent wick changes to help bring the current down. As the current is lowered, the voltage can increase to the set target. For IPG strips other than 11 cm, use the maximum voltages in Table 5.

Different samples focus differently. It may be necessary to focus different samples for a longer or shorter period of time. **Do not focus different sample types or different IPG strip pH ranges in the same focusing tray.**

PROTOCOL 3

Equilibrating the Focused IPG Strips

Once the IPG strips have been focused, it is necessary to equilibrate them in SDS-containing buffers. There are two steps to the equilibration process, a reduction step with equilibration buffer containing TCEP, and an alkylation step with equilibration buffer containing iodoacetamide.

MATERIALS

CAUTION: See Appendix 11 for appropriate handling of materials marked with <!>.

Reagents

Equilibration buffer
  6 M urea <!>
  20% glycerol
  2% (w/v) SDS <!>
  50 mM Tris-Cl (pH 8.8)
   Adjust with H_2O to the final desired volume. Just before use add TCEP <!> and iodoacetamide <!> as per Step 1.
Focused IPG strips (from Protocol 2)

Equipment

Glass tubes

PROCEDURE

1. Prepare two containers each with 30 mL of stock equilibration buffer.

   a. To one container, add 750 µL of 200 mM TCEP.

ANALYSIS OF WHOLE-CELL LYSATES   25

    b. To the other container, add 750 mg of iodoacetamide.

2. Place each strip, **gel side up**, in a glass tube and add 5 mL of the TCEP-containing equilibration solution to each tube. Incubate the strips with gentle agitation for 15 minutes.

3. Decant the solution into a waste bottle and refill the tubes with 5 mL of iodoacetamide-containing equilibration solution. Incubate with gentle agitation for 15 minutes.

PROTOCOL 4

Second Dimension—SDS-PAGE

MATERIALS

CAUTION: See Appendix 11 for appropriate handling of materials marked with <!>.

Reagents

Equilibrated and focused IPG strips (from Protocol 3)
Fixative for SYPRO Ruby or Coomassie Blue staining (40% methanol <!>, 10% acetic acid <!>)
Kaleidoscope prestained standards (Bio-Rad 161-0324)
Running buffer (MES <!> or MOPS <!>)

Equipment

4–12% Bis-Tris SDS-PAGE gels
Gel electrophoresis unit used for the second dimension (Criterion Dodeca cell; Bio-Rad)
Gel packaging tray (Criterion, Bio-Rad)

PROCEDURE

1. Fill the second-dimension chamber with the appropriate 1x running buffer and chill to 4–10°C. The 4–12% Bis-Tris SDS-PAGE gels require MES or MOPS buffer. In the course, we use 1x MES buffer.

2. Remove the comb on the gel cassette and, using a pipette, wash both lanes with running buffer. Position the gel cassette so that it is in an upright position.

   Retain the cassette container for the staining step.

3. Load 5 µL of the Kaleidoscope standards into the single lane.

4. Following equilibration (Protocol 3), remove the IPG strips from the tubes and position them directly on top of the second dimension gel. Use a spatula to be sure that the IPG strip lies evenly on top of the acrylamide gel (Fig. 5).

   Be sure that there are no air bubbles between the IPG strip and the acrylamide gel surface. Orient the IPG strips in the same direction for all samples.

5. Insert the cassettes into the second-dimension electrophoresis unit. Run the gels at 4°C, typically in a cold room (Fig. 6). Alternatively, external cooling units can be purchased.

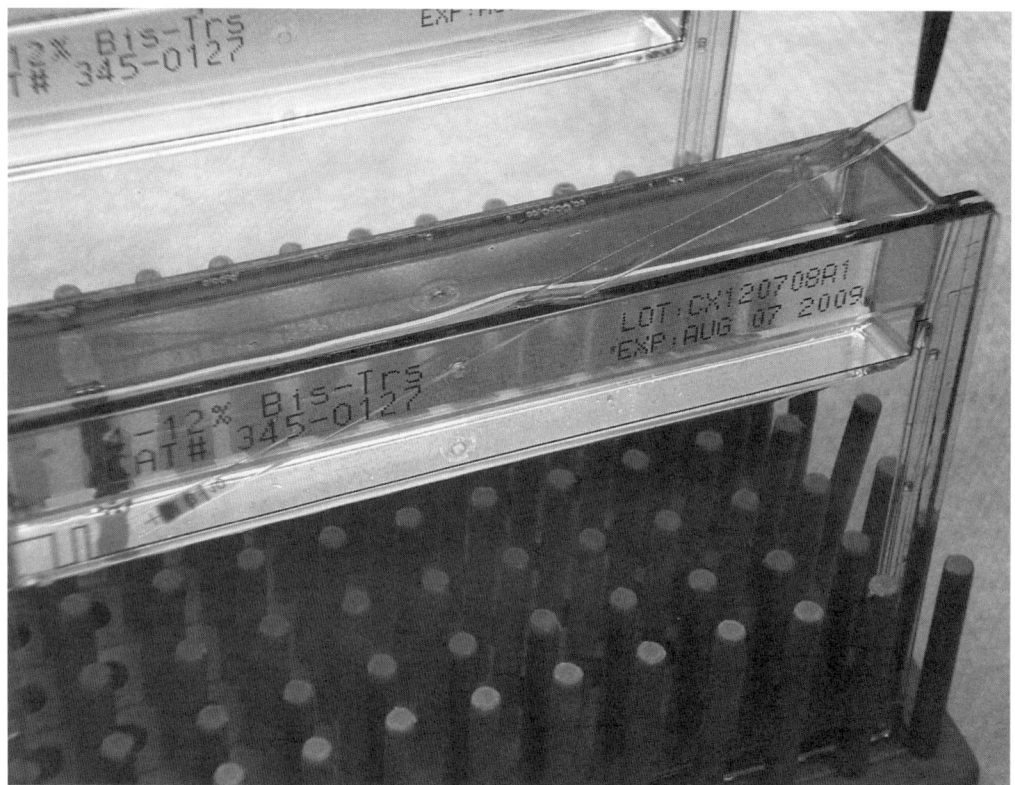

FIGURE 5. After equilibration, the focused IPG strip is placed on top of the second dimension SDS-PAGE gel. The IPG strip sits evenly on the surface and no air bubbles are between the strip and the second dimension gel.

FIGURE 6. The second dimension runs in the Criterion Dodeca Cell which can hold up to 12 gels. The gels are run at a constant voltage of 200 V until the dye line reaches within 1 cm of the bottom of the gel.

6. Program the power supply to run at a constant voltage of 200 V (approximately 1–1.5 hours.). Run the second dimension.

7. After the second dimension is complete, remove the gels from the unit and "crack" the ends of the cassettes to open them. Gently remove the gels from the cassettes and place the gels into labeled containers. Briefly wash the gels for a few minutes in H$_2$O to remove the methanol and acetic acid.

8. Place ~100 mL of fixative solution into the gel packaging tray.

   From this point on, treat the gels with great care to ensure that they are not torn. Always wear gloves when handling the gels to avoid contaminating them with keratin.

9. Allow the gels to fix with gentle agitation for at least 2 hours. Gels can be fixed overnight or for several days if need be.

VISUALIZATION OF SEPARATED PROTEINS

The following protocols provide several methods for staining gel-bound proteins. Protocols 5, 6, and 7 utilize multiplex (fluorescent) stains for total protein, phosphoproteins, and glycoproteins. Multiplex stains are sensitive to light and must be used and stored in the dark. A rapid Coomassie Blue stain and a mass-spectrometer compatible silver stain are also included.

PROTOCOL 5

Total Protein Staining with SYPRO Ruby

The SYPRO Ruby gel stain is recommended in place of conventional silver staining. The sensitivity of Sypro Ruby is comparable to silver staining, quantitation can be achieved with this staining method, and it is compatible with mass spectrometry. Coomassie staining or silver staining can be done after staining with SYPRO Ruby, if desired. Gels can be fixed overnight or for several days if need be.

MATERIALS

CAUTION: See Appendix 11 for appropriate handling of materials marked with <!>.

Reagents

2D gel with separated protein samples
10% Methanol <!>
40% Methanol, 10% acetic acid <!>
SYPRO Ruby stain <!> (Bio-Rad 170-3138)

Equipment

Tray shaker
UV transilluminator

PROCEDURE

1. Immerse the gel in ~100 mL of 40% methanol, 10% acetic acid and gently agitate for a minimum of 2 hours for fixation.
2. Decant the fixation solution.
3. Pour ~200 mL of SYPRO Ruby stain over the gels (use enough stain to cover the gels).
4. Stain for a minimum of 90 minutes.
5. Wash the gels in 10% methanol for a minimum of 1 hour.
6. Use a UV transilluminator (or other suitable device) to visualize the proteins in the gel.

PROTOCOL 6

Phosphoprotein Staining with Pro-Q Diamond

MATERIALS

CAUTION: See Appendix 11 for appropriate handling of materials marked with <!>.

Reagents

2D gel with separated protein samples
Destain solution (20% acetonitrile <!>, 50 mM sodium acetate [pH 4.0])
50% Methanol <!>, 10% acetic acid <!>
Pro-Q Diamond phosphoprotein gel staining kit (Invitrogen MPP33300)

Equipment

Tray shaker
UV transilluminator

PROCEDURE

1. Immerse the gel in ~100 mL of 50% methanol, 10% acetic acid and gently agitate for a minimum of 30 minutes to fix the gel. Repeat to ensure that all of the SDS is removed from the gel. Gels can be left in fix solution overnight.
2. Wash gel in ~100 mL of H_2O for 30 minutes. Repeat twice more. Residual methanol and acetic acid will interfere with phosphoprotein staining.

3. Stain the gel with Pro-Q Diamond stain for 90 minutes. Incubate in the dark with gentle agitation in 75–100 mL of stain. Staining overnight will result in unacceptably high background.

4. Destain with ~100 mL of destain solution for 30 minutes. Repeat two more times. Protect from light. The signal decreases with extensive destaining.

5. Wash the gel with H_2O for 5 minutes. Repeat. If the background is high, leave the gel in the second wash for 20–30 minutes.

6. Use a UV transilluminator (or other suitable device) to visualize the phosphoproteins in the gel.

PROTOCOL 7

Glycoprotein Staining With Pro-Q Emerald

MATERIALS

CAUTION: See Appendix 11 for appropriate handling of materials marked with <!>.

Reagents

2D gel with separated protein samples
3% Acetic acid <!>
50% Methanol <!>, 10% acetic acid
Pro-Q Emerald staining kit (Invitrogen P21875)
   The kit includes staining reagent, staining buffer, and oxidizing reagent (periodic acid <!>).

Equipment

Tray shaker
UV transilluminator

PROCEDURE

1. Immerse the gel in ~100 mL of 50% methanol, 10% acetic acid and gently agitate for a minimum of 30 minutes to fix the gel. Repeat to ensure that all of the SDS is removed from the gel. Gels can be left in fix solution overnight.

2. Wash the gel in ~100 mL of 3% acetic acid for 10–20 minutes. Repeat.

3. Oxidize the carbohydrates with ~50 mL of oxidizing solution for 30 minutes.

4. Wash the gel in ~100 mL of 3% acetic acid for 10–20 minutes. Repeat twice more.

5. Prepare the stain by diluting 1 mL of stock with 50 mL of staining buffer. Incubate the gel and stain together with gentle agitation for 90–120 minutes in the dark. Do not overstain.

30 EXPERIMENT 1

6. Wash in 3% acetic acid for 15–20 minutes. Repeat.

7. Rinse briefly in H$_2$O before placing on the imager. Pro-Q Emerald 300 has an excitation maximum at 280 nm. Use a UV transilluminator (or other suitable device) to visualize the glycoproteins in the gel.

PROTOCOL 8

Staining Total Protein with Colloidal Coomassie Blue

MATERIALS

CAUTION: See Appendix 11 for appropriate handling of materials marked with <!>.

Reagents

2D gel with separated protein samples
Imperial protein stain (Pierce 24615)
10% Methanol <!> (Optional, see Step 5)

Equipment

Tray shaker

PROCEDURE

1. Place the gel in a clean tray. Wash the gel in 200 mL of H$_2$O with gentle shaking for 15 minutes.

2. Mix the Imperial stain immediately before use to disperse any aggregates.

3. Add enough stain to completely cover the gel.

4. Stain the gel with gentle agitation for 2 hours.

5. Discard the stain and wash with 200 mL of H$_2$O or 200 mL of 10% methanol to reduce background. Wash overnight to obtain the highest sensitivity.

    Replace the H$_2$O multiple times to decrease wash time.

PROTOCOL 9

Staining Total Protein with Mass Spectrometry-compatible Silver Stain

Coomassie stained gels or multiplex stained gels can be restained with this method. Begin with Step 2 for previously fixed gels. The timing of the steps in silver staining is important. Note that silver staining is not a quantitative stain.

MATERIALS

CAUTION: See Appendix 11 for appropriate handling of materials marked with <!>.

Reagents

2D gel with separated protein samples
1.0% Acetic acid <!>
0.2% AgNO$_3$ <!>, 0.08% formaldehyde <!>
35% Ethanol <!>
40% Methanol <!>, 10% acetic acid
6.0% Na$_2$CO$_3$ <!>, 0.05% formaldehyde, 0.0004% Na$_2$S$_2$O$_3$
0.02% Na$_2$S$_2$O$_3$ (pentahydrate)

Equipment

Tray shaker

PROCEDURE

1. Fix the gel in 40% methanol, 10% acetic acid for at least 2 hours or up to overnight.
2. Wash the gel in 35% ethanol for 20 minutes. Repeat two more times.
3. Sensitize the gel in 0.02% Na$_2$S$_2$O$_3$ (pentahydrate) for 2 minutes.
4. Wash the gel in MilliQ-H$_2$O for 5 minutes. Repeat two more times.
5. Stain the gel in 0.2% AgNO$_3$, 0.08% formaldehyde for 20 minutes.
6. Wash the gel in MilliQ-H$_2$O for 1 minute. Repeat.
7. Develop the gel in 6.0% Na$_2$CO$_3$, 0.05% formaldehyde, 0.0004% Na$_2$S$_2$O$_3$.
8. Stop development with 40% methanol, 10% acetic acid for 5 minutes.
9. Store the gel in 1.0% acetic acid at 4°C.

PROTOCOL 10

Image Analysis and Retrieval of Proteins from 2D Gels

MATERIALS

CAUTION: See Appendix 11 for appropriate handling of materials marked with <!>.

Reagents

2D gel, stained
Ethanol <!> (Optional, see Step 7)

Equipment

96-well template
ChemiDoc XRS CCD digital camera (Bio-Rad)
Dermal punch, 1.5-mm or 2.0-mm
Polypropylene digest plate
Spot profile analysis software

PROCEDURE

Gel Imaging and Analysis

1. Image the stained gels using a CCD digital camera according to the manufacturer's protocol. Save the images for analysis and print the image at maximum size.

   Alternatively, stained gels can be imaged using a fluorescent scanner, phosphorimager, or densitometer (Fig. 7).

2. Compare spot profiles using either Flicker (free program available through download on the internet) or PDQuest gel analysis software (Bio-Rad) (Figs. 8 and 9).

3. After reviewing the 2D gels, note any differences and/or changes of expression between the glucose (fermentation) and glycerol (respiration) gels. Annotate the gel images and print out a copy.

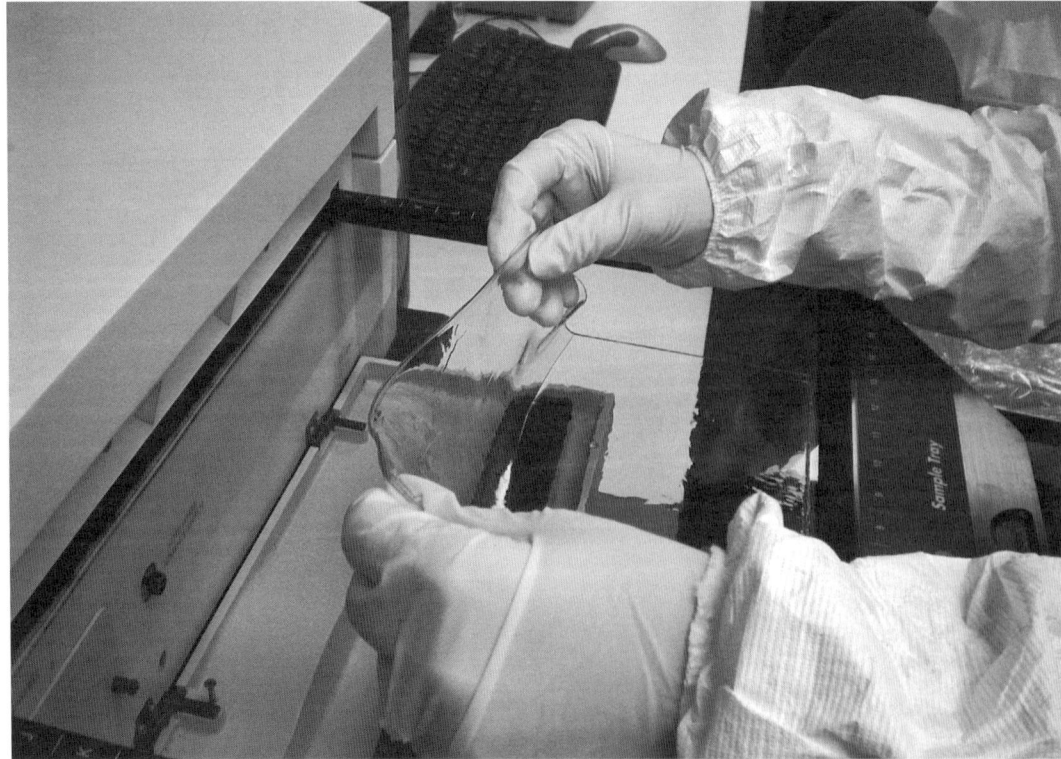

FIGURE 7. A fluorescent scanner (Bio-Rad Molecular Imager FX) for imaging a stained gel.

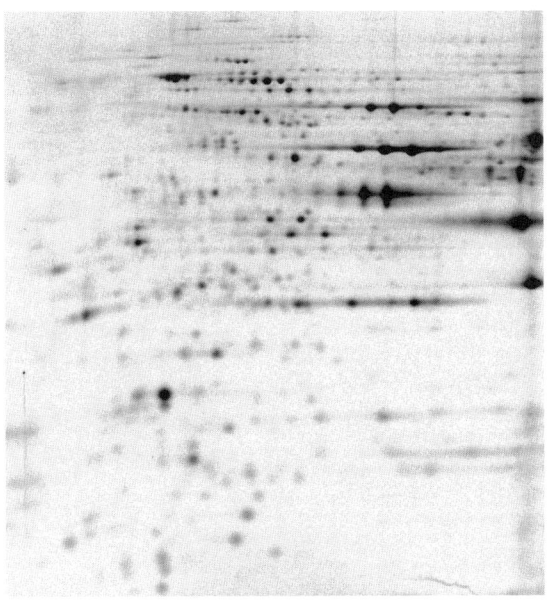
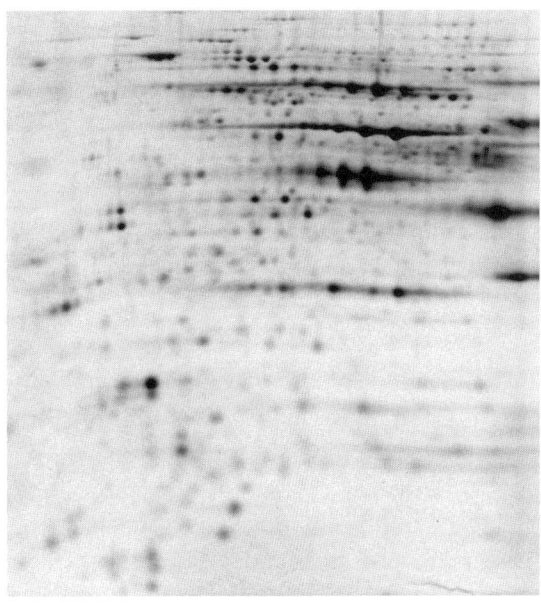

FIGURE 8. An example of a large format glucose (fermentation) 2D gel. The IPG strip is pH 4–7 and is 18 cm in length. The second dimension gel is a 12.5% homogenous acrylamide gel and the dimensions are 20 x 20 cm. The gel stain is SYPRO Ruby.

FIGURE 9. An example of a large format glycerol (respiration) 2D gel. The IPG strip is pH 4–7 and is 18 cm in length. The second dimension gel is a 12.5% homogenous acrylamide gel and the dimensions are 20 x 20 cm. The gel stain is SYPRO Ruby.

Spot Cutting

4. Using the annotated printed image of the desired gel and a 96-well template as a record, label the wells that will be designated for each spot.

5. Pipette 50 μL of ultra pure H_2O into the appropriate number of wells in a polypropylene digest plate.

6. Cut the interesting protein spots from the gel using a 1.5 mm or 2.0 mm dermal punch and dispense gel plugs into the plate provided. Place a maximum of 2 plugs in each well. For a dark spot, one plug is sufficient.

7. Rinse the dermal punch with H_2O or ethanol between cuts to avoid cross contamination.

General considerations when performing tryptic digests

- Always use the highest grade of reagents available, i.e., Optima grade or HPLC grade H_2O and acetonitrile.
- Always make the reagents fresh prior to using.
- Some plastic ware can have contaminants from manufacturing; consult your local facility for recommendations.
- Polystyrene is incompatible with organics; use only polypropylene.
- Clean all work areas prior to working, including racks. Work in a hood when possible.
- Do not wear wool or silk clothing.

PROTOCOL 11

In-gel Tryptic Digestion

MATERIALS

CAUTION: See Appendix 11 for appropriate handling of materials marked with <!>.

Reagents

Alkylation solution (55 mM iodoacetamide) <!>
    Dissolve 102 mg of IAA to a final volume of 10 mL with 100 mM ammonium bicarbonate.
100 mM Ammonium bicarbonate
    Dissolve 0.395 g of NH_4HCO_3 made up to 50 mL with HPLC grade H_2O.
Dehydration solution (HPLC grade acetonitrile) <!>
Extraction solution (1% formic acid <!>, 2% acetonitrile)
    Mix 100 µL of formic acid and 200 µL of acetonitrile with 10 mL of HPLC grade H_2O.
Protein samples embedded in gel plugs (from Protocol 10)
Reduction solution (10 mM DTT) <!>
    Dissolve 15.4 mg of DTT to a final volume of 10 mL with 100 mM ammonium bicarbonate.
Trypsin solution
    Dissolve 100 µg of lyophilized trypsin <!> in 1000 µL of reconstitution buffer (50 mM acetic acid <!>) provided with the trypsin. Aliquot the trypsin into 200-µL aliquots and freeze at –80°C. The trypsin is inactive at or below pH 4.

Equipment

Incubator set at 37°C
PCR plate

PROCEDURE

1. Remove any excess solution from each well ensuring that the gel plug remains in the well.

2. If the gel has been stained with Coomassie Blue or SYPRO Ruby, destain the plugs by dispensing 50 µL of 50 mM ammonium bicarbonate and 50 µL of acetonitrile into each well. Incubate for 10 minutes at 37°C. Aspirate and discard the solution from each well. Repeat the process.

3. Dehydrate the gel plugs by dispensing 50 µL of acetonitrile into each well and incubating them for 5 minutes at 37°C. Aspirate the solution from each well. Remove any residual acetonitrile by incubation for 10 minutes at 37°C.

4. To reduce the proteins, dispense 50 µL of 10 mM DTT into each well and incubate for 20 minutes at 37°C . Aspirate the solution from each well.

5. To alkylate the proteins, add 30 µL of 55 mM iodoacetamide into each well and incubate for 20 minutes at 37°C. Aspirate the solution from each well.

6. Add 50 µL of acetonitrile into each well and incubate for 5 minutes at 37°C. Aspirate the

solution from each well. Repeat the process. Remove any residual acetonitrile by incubation for a further 5 minutes at 37°C.

7. Activate the trypsin solution as follows. Add 800 μL of 50 mM ammonium bicarbonate to a 200-μL aliquot of trypsin (aliquot is 100 μg/mL). Thus, 10 μL will contain 200 ng of trypsin (20 ng/μL).

    The addition of ammonium bicarbonate activates the trypsin, which is maximally active in the pH range 7–9.

8. Add 15 μL of the active trypsin solution to each well and incubate for 10 minutes at room temperature to allow the trypsin to absorb into the gel plugs.

9. Add 15 μL of 50 mM ammonium bicarbonate to each well and incubate for 4 hours at 37°C (incubate overnight if needed).

10. Dispense 20 μL of extraction solution into each well and incubate for 30 minutes at room temperature.

11. Aspirate the solution from each well and transfer it to a new PCR plate.

PROTOCOL 12

MALDI Plate Spotting

MALDI target plates are available in 100-well or 192-well formats. The MALDI TOF/TOF mass spectrometer used in the course is configured for the 192-well format, and the plate does not require a backing.

MATERIALS

CAUTION: See Appendix 11 for appropriate handling of materials marked with <!>.

Reagents

Acetonitrile <!>
Calibration/matrix mix
Formic acid <!>
Methanol <!>
Protein samples
Spotting matrix (α-cyano-4-hydroxycinnamic acid) <!>
    10 mg/mL of α-cyano-4-hydroxycinnamic acid in 50% acetonitrile (HPLC grade), 50% H_2O (HPLC grade) and 0.1% TFA <!>. Add 2 mg/mL ammonium citrate just before use.

Equipment

Beaker
MALDI target plate
Sonicator <!>

PROCEDURE

Target Plate Cleaning

1. Rinse a target plate with methanol and wipe it with a Kimwipe.
2. Place the plate in a beaker and cover it with methanol. Add a drop of formic acid. Sonicate for 30 minutes.
3. Remove the plate and wipe off the methanol/formic acid.
4. Place the plate in a beaker and cover it with acetonitrile. Sonicate for 30 minutes.
5. Wipe the plate dry.

Spotting Matrix and Sample onto a Target Plate

Samples for mass spectrometry analysis need to be free of salts and low-level contaminants. Samples from in-gel digests can usually be spotted directly onto target plates. There are numerous matrices that can be used; one of the more common matrices is α-cyano-4-hydroxycinnamic acid.

6. Fill out a sample sheet for TOF/TOF plates (192-well format). It is necessary to keep a record of which sample spot goes to which spot on the target plate.
7. Apply 0.5 µL of sample to a well on the target plate. Do not let the sample dry completely before applying matrix (Fig. 10).
8. Apply 0.5 µL of matrix directly on top of the sample. The sample and matrix will cocrystallize. Allow them to air dry. Dried matrix should have a uniform appearance and be off-white to slightly yellow in color.
9. Apply 0.5 µL of the calibration/matrix mix to various wells throughout the plate.
10. Insert the target plate into the mass spectrometer and run according to Experiment 3.
11. Analyze the mass spectrometry results (see Experiment 8).

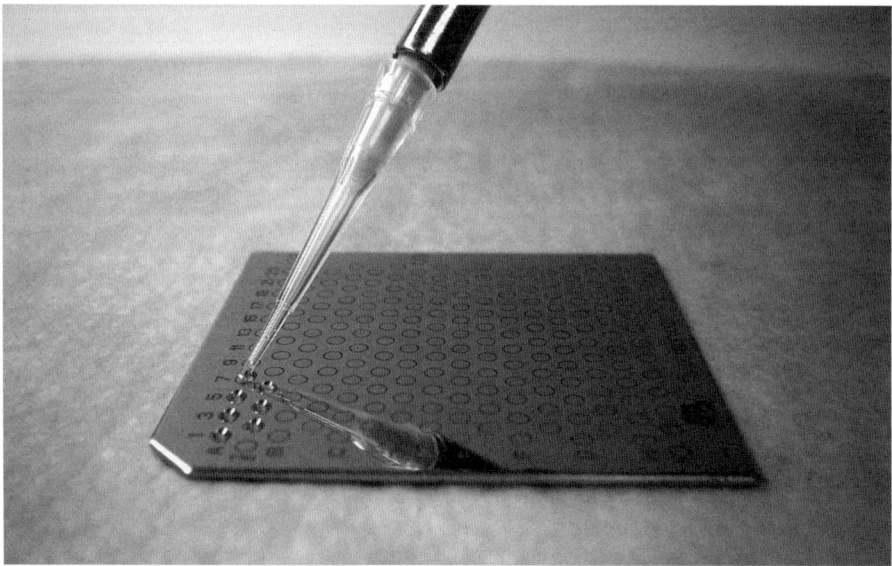

FIGURE 10. The sample is spotted onto a single well of a MALDI target and the matrix is then spotted directly on top of the sample. The sample and matrix will cocrystallize as they dry.

EXPERIMENT 2

Purification of Protein Complexes for Mass Spectrometry Analysis

Protein interactions are essential for nearly all biological processes. Knowing protein–protein interactions is a key element in understanding the function of a protein in the cell. Knowledge of protein interactions improves our understanding of diseases and provides putative targets for therapeutic intervention. The principle of "guilt by association" can be used to infer biochemical function of uncharacterized proteins based on the protein partners with which they interact. For characterized proteins, unexpected protein interactions may provide evidence for a novel function(s).

To identify protein–protein interactions, one approach is the biochemical purification of a target protein from cells or tissues under nondenaturing conditions followed by the mass spectrometric identification of the components of the purified protein complex. The combination of highly specific protein purification strategies and mass spectrometry has proven to be a very successful approach for identifying protein–protein interactions. Two important issues need to be considered. One issue is the purification of protein complexes from cells or tissues in sufficient quantity for mass spectrometry analysis. Second, the purified complexes need to be isolated efficiently with a minimum amount of nonspecific interactions. Investigators could spend a considerable amount of time and resources trying to confirm nonspecific interactions. The goal is to isolate sufficient quantity of the intact complexes using as few steps as possible. The purification conditions must balance stabilizing protein interactions among the physiological components while avoiding nonspecific or nonphysiological interactions with other proteins.

This experiment teaches two powerful affinity strategies for isolating protein complexes from yeast. The first experiment uses tandem affinity purification (TAP) to isolate protein complexes from yeast cells. In the second experiment, students are trained to use immunopurification of protein complexes using magnetic beads to rapidly isolate protein complexes from yeast. For training students in the art of protein purification, isolating protein complexes from yeast strains with affinity-tagged genes provides a convenient and practical demonstration of the fundamental techniques and methods. The approaches described in this experiment have been applied to other organisms including prokaryotes and higher eukaryotes. When teaching students to purify protein complexes using these two protocols, we emphasize three critical concepts. First, work quickly to minimize the dissociation of endogenous complexes and to limit the formation of nonphysiological interactions. Second, work at 4°C and keep solutions on ice to limit unwanted proteolysis and degradation. Third, minimize diluting protein extracts to minimize the dissociation of native complexes and formation of nonspecific interactions. Most protein complexes utilize noncovalent forces to maintain the interactions.

TANDEM AFFINITY PURIFICATION (TAP) OF PROTEIN COMPLEXES

Contributed by Andrew J. Link, Connie Weaver, and Adam Farley *(Vanderbilt University School of Medicine, Nashville, Tennessee 37232)*.

Affinity purification of protein complexes begins with the purification of a tagged protein under native conditions. Mass spectrometry is then used to identify all of the copurifying proteins. For the purification step, epitope tags that allow affinity purification are ideal because relatively pure protein complexes can be obtained in only one or two steps, thus minimizing the loss of associated proteins. In yeast, an innovation in the epitope-tagging procedure uses a TAP tag (Rigaut et al. 1999). The IgG-binding domain of Protein A from *Staphylococcus aureus* (ProtA) and the calmodulin binding peptide (CBP) are fused in tandem and separated by a tobacco etch virus (TEV) protease cleavage site, allowing for affinity purification based on both epitopes (Fig. 1). Rapid PCR-based strategies have been developed to introduce epitope tags at yeast chromosomal loci. The TAP affinity tag and purification method allows efficient recovery of proteins present at low cellular concentrations under native conditions (Fig. 2). Expressing the target protein at its natural levels avoids the assembly of overexpressed proteins into nonphysiological complexes.

In this section, we purify TAP-tagged complexes from *Saccharomyces cerevisiae* cells that have already been constructed and the cell pellets generated. As a control, we will purify in parallel the parental strain of the TAP-tagged yeast strain in order to measure the background noise. A portion of the purified protein complexes will be run on a silver-stained SDS-PAGE gel and the rest will be analyzed by mass spectrometry.

Following this section on TAP, the immunopurification of protein complexes using magnetic beads is described.

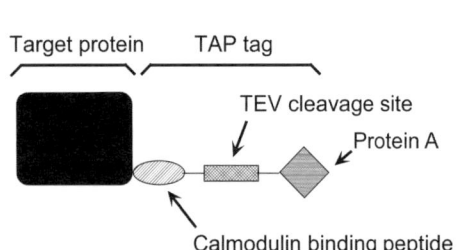

FIGURE 1. Structure of the tandem affinity purification (TAP) tag. The various components of the TAP tag fused at the carboxyl terminus of the target protein are shown. The IgG-binding domain of Protein A from *Staphylococcus aureus* (ProtA) and the calmodulin binding peptide (CBP) are fused in tandem and separated by a TEV protease cleavage site, allowing for affinity purification based on both epitopes.

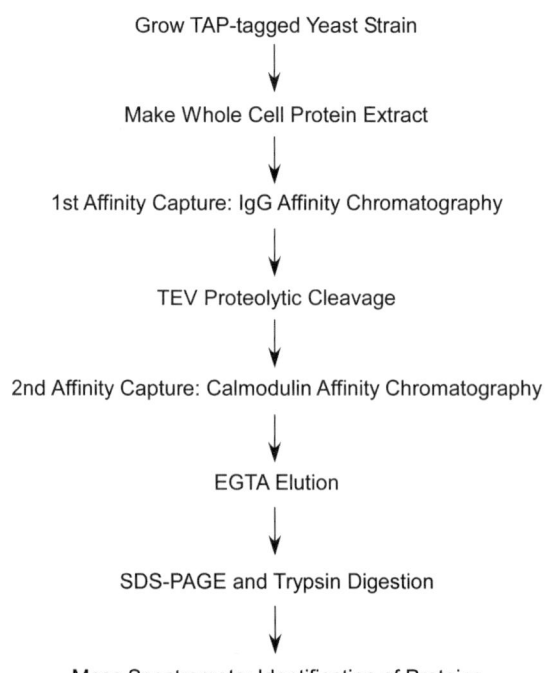

FIGURE 2. Experimental flow chart of the tandem affinity purification protocol.

PROTOCOL 1

Growing and Harvesting TAP-tagged Yeast Cells

MATERIALS

CAUTION: See Appendix 11 for appropriate handling of materials marked with <!>.

Reagents

Drop-out mix (lacking Histidine)

Adenine	0.5 g	Leucine	10.0 g
Alanine <!>	2.0 g	Lysine	2.0 g
Arginine	2.0 g	Methionine	2.0 g
Asparagine	2.0 g	para-Aminobenzoic acid	0.2 g
Aspartic acid	2.0 g	Phenylalanine	2.0 g
Cysteine	2.0 g	Proline	2.0 g
Glutamine	2.0 g	Serine	2.0 g
Glutamic acid	2.0 g	Threonine	2.0 g
Glycine	2.0 g	Tryptophan	2.0 g
Histidine	—	Tyrosine	2.0 g
Inositol	2.0 g	Uracil	2.0 g
Isoleucine	2.0 g	Valine	2.0 g

SC–his agar plates

Prepare 1 liter of media in a 2-liter flask.

Component	Amount to Add	Final Concentration
Bacto-yeast nitrogen base without amino acids*	6.7 g	0.67%
Bacto-agar	20 g	2%
Glucose	20 g	2%
Drop-out mix minus Histidine	2 g	0.2%

*Yeast nitrogen base without amino acids (YNB) is sold either with or without ammonium sulfate. This recipe is for YNB with ammonium sulfate. If the bottle of YNB is lacking ammonium sulfate, add 5 g of ammonium sulfate and only 1.7 g of YNB.

TAP-tagged yeast strains (Invitrogen)

Two groups have made TAP-tagged yeast strain libraries for most genes in the yeast genome (Ghaemmaghami et al. 2003; Gavin et al. 2006). The entire library of TAP strains and individual TAP strains are commercially available (Invitrogen). For teaching purposes, we use TAP-tagged yeast genes that are highly expressed and known components in stable complexes. The TAP-tagged genes *Rpg1* (eIF3), *Gcd11* (eIF2), *Gcd6* (eIF2B), and *Nup84* (nuclear pore complex) are components in different translation initiation or RNA transport complexes and have been successfully purified by students in the CSHL Proteomics course.

YPD agar plates

To prepare YPD plates, add 20 g of Bacto-agar (2%) to 1 liter of YPD liquid media in a 2-liter flask before autoclaving. Fill sterile Petri dishes with 30–40 mL of autoclaved media.

YPD liquid medium

Prepare 1 liter of media in a 2-liter flask.

40 EXPERIMENT 2

Component	Amount to Add	Final Concentration
Bacto-yeast extract	10 g	1%
Bacto-peptone	20 g	2%
Dextrose	20 g	2%

Dissolve ingredients in deionized H_2O. Autoclave for 15 minutes at 250°F (121°C) and 15 psi.

Equipment

Centrifuge, low speed
Centrifuge bottles, 250-mL and 1-liter
Flasks, 2-liter
Gravity incubator set at 30°C
Shaking incubator set at 30°C

PROCEDURE

1. Streak out the frozen TAP-tagged yeast and control strains onto SC–his and YPD agar plates, respectively, and grow at 30°C in a gravity incubator until colonies appear.

   The O'Shea and Weissman yeast TAP strains that are distributed by Invitrogen are tagged with a HIS3 selectable marker. Other TAP strains are typically tagged with a Kan marker (G418 resistance).

2. Inoculate TAP-tagged and control yeast strains into 10 mL of YPD medium and grow overnight at 30°C.

3. For each liter of YPD, inoculate with 2 mL of the overnight culture.

   For TAP-tagged genes associated with the translation initiation complexes, we grow 2 L of each TAP-tagged strain.

4. Grow the cultures for 12–18 hours at 30°C with shaking to an OD_{600} of 2–4.

   Do not let cultures grow to stationary phase. Yeasts grown to stationary phase are very difficult to lyse.

5. Centrifuge the cells at 2300g for 10 minutes at 4°C.

6. Wash the cells with 300 mL of ice-cold deionized H_2O. Combine the cell suspensions equally into two large centrifuge bottles.

7. Centrifuge at 2300g for 10 minutes at 4°C.

8. Resuspend each pellet in 200 mL of ice-cold dH_2O and transfer to a preweighed 250-mL centrifuge bottle.

9. Centrifuge at 2300g for 10 minutes at 4°C.

10. Pour off the supernatant and weigh the cell pellet.

    Cells can be frozen (–80°C) for several months with no adverse affects. For teaching purposes in this course, yeast cell pellets are prepared for the students ahead of time.

PROTOCOL 2

Making Yeast Cell Extracts for Purifying TAP-tagged Complexes

MATERIALS

CAUTION: See Appendix 11 for appropriate handling of materials marked with <!>.

Reagents

NP-40 buffer

To prepare 3 liters:

Component	Amount to Add	Final Concentration
Na_2HPO_4 (FW 142)	2.56 g	6 mM
$NaH_2PO_4 \cdot H_2O$ (FW 138)	1.66 g	4 mM
100% NP-40	30 mL	1%
NaCl	26.3 g	150 mM
0.5 M EDTA	12 mL	2 mM
NaF <!>	6.3 g	50 mM
Leupeptin <!>	12 mg	4 µg/µL
100 mM Na_3VO_4 <!>	3 mL	0.1 mM

Bring to a final volume of 3 liters with distilled H_2O. Dispense into 500-mL aliquots. Just prior to use, add 10 COMPLETE EDTA-free protease inhibitor tablets (Roche), 1.3 mL of 0.5 M benzamidine, and 5.0 mL of 0.1 M PMSF <!> to 500 mL of NP-40 buffer.

Yeast, TAP-tagged and control cell pellets

Equipment

BeadBeater with stainless-steel lysis chamber and ice-H_2O chamber (Biospec Products)
Centrifuge
Centrifuge tubes, 250-mL conical
Glass beads, 0.4–0.6-mm acid-washed (Biospec Products)

PROCEDURE

1. Fill the BeadBeater's stainless-steel chamber half-full with ice-cold 0.4–0.6-mm glass beads. Add either the fresh yeast cell pellets or partially thawed cells. Fill the chamber to the neck with additional beads (Fig.3A). Add ~100 mL of ice-cold NP-40 buffer to completely cover the beads (Fig. 3B). Insert the rotor assembly along with the gray rubber gasket and screw on the ice-H_2O chamber (hold the chamber upside down) (Fig. 3C–E).

    Prior to starting the experiment, chill all buffers, the metal chamber, and the glass beads on ice. Before attaching the rotor, minimize the amount of air in the chamber by adding additional NP-40 buffer. Avoid getting beads on the lip and threads of the BeadBeater metal chamber to prevent leaks.

2. Add crushed ice to the ice-water jacket surrounding the BeadBeater chamber (Fig. 3F).

3. Lyse the cells with 10 cycles of 30 seconds on and 30 seconds off.

    The BeadBeater will disrupt over 90% of the cells. The homogenization procedure involves cell crushing or "cracking" action rather than high shear.

FIGURE 3. (A) Level of silica beads in the BeadBeater's stainless steel grinding chamber. (B) Level of cells added to grinding chamber. (C) Inserting the rotor into the grinding chamber. (D) Attaching the grinding chamber to the ice jacket. (E) Grinding chamber and ice-water chamber screwed together. (F) With ice-water mixture added, the BeadBeater is ready to grind yeast cells to make crude cell extracts.

4. Carefully transfer the crude lysate to a 250-mL conical centrifuge tube (Fig. 4).

    As much as is possible, avoid transferring glass beads.

5. Wash the beads with an additional 100 mL of NP-40 lysis buffer and add the wash to the crude lysate.

6. Centrifuge the lysate at 2300g for 5 minutes at 4°C.

7. Transfer the cleared lysate to a fresh 250-mL conical centrifugation tube on ice.

    It is critical to proceed immediately to the first affinity capture step using IgG-Sepharose beads.

FIGURE 4. Transferring crude cell extract from the BeadBeater's lysis chamber to a 250-mL centrifuge tube.

PROTOCOL 3

First Affinity Enrichment Step: IgG Affinity Capture of TAP-tagged Complexes from Cell Extracts

MATERIALS

CAUTION: See Appendix 11 for appropriate handling of materials marked with <!>.

Reagents

IgG-Sepharose 6 Fast Flow resin (GE Healthcare)

To prepare the IgG-Sepharose, wash the IgG-Sepharose beads with NP-40 buffer without protease inhibitors and resuspend in NP-40 buffer to make a 1:1 slurry.

IPP150 buffer

To prepare 100 mL:

Component	Amount to Add	Final Concentration
1 M Tris-Cl (pH 8.0)	1 mL	10 mM
5 M NaCl	3 mL	150 mM
10% NP-40	1 mL	0.1%

Adjust to a final volume of 100 mL with distilled H_2O.

NP-40 buffer

To prepare 3 liters:

Component	Amount to Add	Final Concentration
Na_2HPO_4 (FW 142)	2.56 g	6 mM
$NaH_2PO_4 \cdot H_2O$ (FW 138)	1.66 g	4 mM
100% NP-40	30 mL	1%
NaCl	26.3 g	150 mM
0.5 M EDTA	12 mL	2 mM
NaF <!>	6.3 g	50 mM
Leupeptin <!>	12 mg	4 µg/µL
100 mM Na_3VO_4 <!>	3 mL	0.1 mM

Bring to a final volume of 3 liters with distilled H_2O. Dispense into 500-mL aliquots. Just prior to use, add 10 COMPLETE EDTA-free protease inhibitor tablets (Roche), 1.3 mL of 0.5 M benzamidine, and 5.0 mL of 0.1 M PMSF <!> to 500 mL of NP-40 buffer.

TEVCB buffer

To prepare 50 mL:

Component	Amount to Add	Final Concentration
1 M Tris-Cl (pH 8.0)	0.5 mL	10 mM
5 M NaCl	1.5 mL	150 mM
10% NP-40	0.5 mL	0.1%
0.5 M EDTA	50 µL	0.5 mM
1 M DTT <!>	25 µL	1.0 mM

Adjust to a final volume of 50 mL with distilled H_2O. Add the DTT just prior to use.

Tobacco etch virus protease (Invitrogen AcTEV 12575-015)

Yeast cell lysate (prepared in Protocol 2, Step 7)

Equipment

Chromatography column, 0.8 cm x 4 cm disposable (Poly-Prep, Bio-Rad)
Centrifuge
Rotator
Reservoir for chromatography column

PROCEDURE

1. Add 1000 µL of IgG-Sepharose beads in NP-40 buffer (1:1 slurry) to the cleared lysate (from Protocol 2, Step 7) and incubate for 1 hour on a rotator at 4°C (Fig. 5).

   Alternatively, a nutator can be used to incubate and gently agitate the IgG beads with the cell lysate.

2. Centrifuge the mixture of lysate and IgG beads at 200g for 2 minutes at 4°C. Slowly pour off the lysate, trying not to disturb the IgG beads.

3. Add 30 mL of ice-cold IPP150 buffer to the beads.

4. Pour the slurry of IPP150 and IgG beads into a disposable chromatography column using a reservoir (Fig. 6). Wash the 250-mL centrifuge tube with 10 mL of IPP150 buffer and add to the column reservoir. Allow the column to pack at 4°C.

   Watch the column to ensure that the IgG bead bed does not go dry, which can significantly lower the final recovery of protein complexes from the resin.

5. Wash the beads by slowly adding 10 mL of ice-cold TEVCB buffer to the column.

6. Cap the bottom of the column and add 1 mL of TEVCB containing 300 units of TEV protease. Cap the top of the column. Incubate for 1 hour on a rotator at room temperature (Fig. 7).

   The TEV enzyme has maximum catalytic activity near room temperature. Alternatively, a nutator can be used to incubate and gently agitate the IgG beads with the TEV protease.

FIGURE 5. Incubating the cell extract with IgG-Sepharose beads on a rotator to capture protein complexes in the first affinity capture.

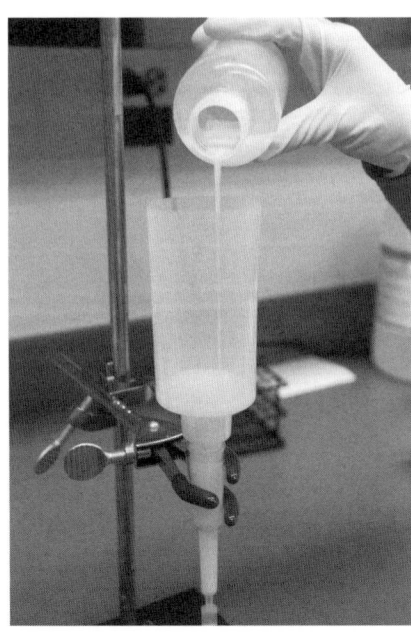

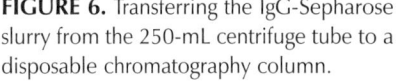

FIGURE 6. Transferring the IgG-Sepharose slurry from the 250-mL centrifuge tube to a disposable chromatography column.

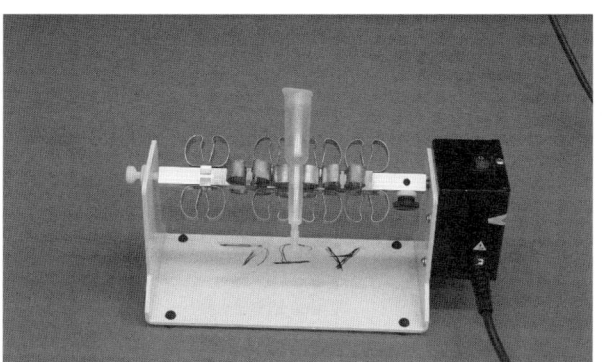

FIGURE 7. Incubating the IgG-Sepharose slurry with TEV protease in a capped chromatography column on a rotator.

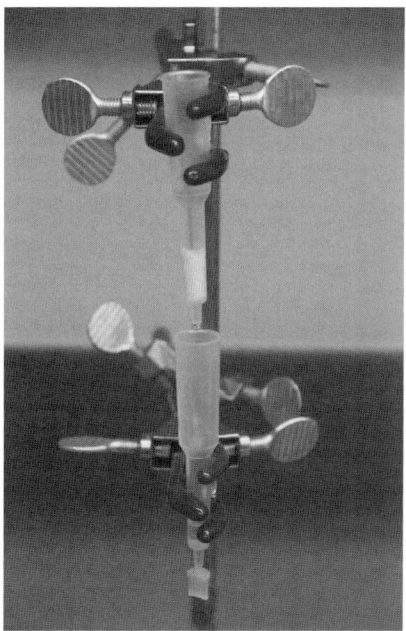

FIGURE 8. Transferring the TEV-cleaved complexes from the first affinity capture to a new chromatography column.

7. Drain the eluate into a new disposable chromatography column that is sealed at the bottom (Fig. 8). Wash the IgG-Sepharose column with 1 mL of TEVCB and combine into the new column.

   It is important to proceed immediately to the second affinity capture step, using calmodulin affinity resin, to minimize the dissociation of the native complexes and limit nonspecific interactions.

PROTOCOL 4

Second Affinity Enrichment Step: Calmodulin Affinity Capture of TAP-tagged Complexes

MATERIALS

CAUTION: See Appendix 11 for appropriate handling of materials marked with <!>.

Reagents

Anti-TAP antibody (Open Biosystems CAB1001; see Step 5)

1 M $CaCl_2$ <!>

Calmodulin affinity resin (Stratagene 214303-52)

   To prepare the calmodulin affinity beads, wash the affinity beads with 0.1% CBB buffer and resuspend in 0.1% CBB buffer minus β-mercaptoethanol to make a 1:1 slurry.

CBB buffer (0.1% and 0.02%)

To prepare 100 mL:

Component	Amount to Add	Final Concentration
10% NP-40	1 mL	0.1%
1 M Tris-HCl (pH 8.0)	1 mL	10 mM
5 M NaCl	3 mL	150 mM
1 M Magnesium acetate	100 µL	1 mM
1 M Imidazole <!>	100 µL	1 mM
1 M CaCl$_2$ <!>	200 µL	2 mM
14.3 M β-Mercaptoethanol <!>	70 µL	10 mM

Adjust to a final volume of 100 mL with distilled H$_2$O. Add the β-ME to the CBB buffer just prior to use. The 0.02% CBB buffer is identical to 0.1% CBB buffer except 200 µL of 10% NP-40 is used.

CEB buffer

To prepare 10 mL:

Component	Amount to Add	Final Concentration
1 M Tris-HCl (pH 8.0)	0.1 mL	10 mM
5 M NaCl	0.3 mL	150 mM
1 M Imidazole	10 µL	1 mM
1 M Magnesium acetate	10 µL	1 mM
0.5 M EGTA	400 µL	20 mM
14.3 M β-Mercaptoethanol	7 µL	10 mM

Adjust to a final volume of 10 mL with distilled H$_2$O. Add the β-ME to the CBB buffer just prior to use.

50 mM DTT <!>
IgG elution (collected in the Poly-Prep column in Protocol 3, Step 7)
100 mM Iodoacetamide <!>
1 M Tris (pH 8.0)
Trypsin <!>, sequencing grade (Promega)

Equipment

Incubators set at 30°C, 37°C, and 65°C
Microcentrifuge tubes, 1.5-mL
Rotator
Vacuum evaporator (e.g., SpeedVac)

PROCEDURE

1. Add 6 mL of 0.1% CBB to the IgG elution that was collected in the Poly-Prep column (Protocol 3, Step 7).

    Ca^{2+} ions are required for binding of the CBP to calmodulin.

2. Add 300 µL of calmodulin affinity beads in 0.1% CBB buffer (1:1 slurry). Incubate for 1 hour at 4°C on the rotator.

    Alternatively, a nutator can be used to incubate and gently agitate the CBB beads with the IgG elution.

3. Open the column and drain. Wash the column 2x with 20 mL of 0.02% CBB buffer.

   The 0.02% CBB wash reduces the amount of NP-40 detergent in the final elution. NP-40 detergent interferes with the mass spectrometric analysis. The β-mercaptoethanol is added to the CBB buffer just prior to use.

4. Elute the protein complexes with 1 mL of CEB into a 1.5-mL microcentrifuge tube (Fig. 9).

   EGTA in the CEB buffer is used to chelate the Ca^{2+} ions and subsequently elute the proteins from the calmodulin affinity beads.

5. Remove 5–100 μL of the elution and save it for SDS-PAGE analysis to determine the complexity of the sample and the relative amounts of proteins obtained.

   An aliquot of the elution is mixed with Laemmli loading buffer and analyzed by SDS-PAGE. The different protein fractions from the TAP protocol can be monitored using SDS-PAGE and Western analysis using an anti-TAP antibody.

6. Add 100 μL of 1 M Tris (pH 8.0) to the remaining volume of elution.

   To bring the solution to the pH range where trypsin has maximal activity, 1 M Tris (pH 8) is added. Avoid precipitating captured protein complexes because re-solubilizing precipitated complexes is often very difficult.

7. Add 50 μL of 1 M $CaCl_2$.

   An excess of $CaCl_2$ is added to neutralize the EGTA in the sample, because EGTA will chelate divalent metal ions. This is key, because for trypsin to be active, divalent metal ions must be present.

8. Add 0.1 volume of 50 mM DTT and incubate at 65°C for 10 minutes.

   DTT is used to reduce the disulfide bonds between cysteine residues.

9. Add 0.1 volume of 100 mM iodoacetamide (IAA) and place at 30°C for 30 minutes in the dark.

   IAA is used to alkylate the cysteine residues to prevent the formation of disulfide bonds.

10. Add modified sequencing-grade trypsin (1 μg/μL) at a ratio of 50:1 substrate to trypsin. Incubate the reaction overnight at 37°C.

    The amount of substrate present is based upon the absorbance of the elution (from Step 4) at 280 nm. Use CEB buffer as a blank.

11. Analyze the trypsin-digested proteins as described in Experiment 4 or Experiment 6.

    Use a reversed-phase-trapping cartridge to capture the peptides, desalt the sample, and concentrate the peptides prior to mass spectrometry analysis (see Appendix 8).

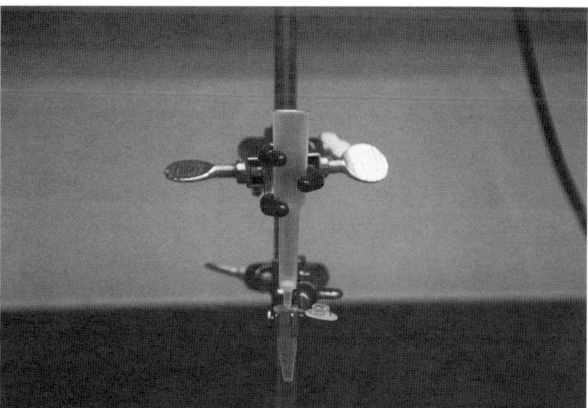

FIGURE 9. Eluting protein complexes from the calmodulin affinity beads into a 1.5-mL microcentrifuge tube.

AFFINITY PURIFICATION OF PROTEIN COMPLEXES

Contributed by Ileana M. Cristea (*Department of Molecular Biology, Princeton University, Princeton, New Jersey 08544*) and Brian T. Chait (*Laboratory of Mass Spectrometry and Gaseous Ion Chemistry, Rockefeller University, New York, New York 10065*).

Knowledge of the composition of protein complexes provides key insights into their functions. A variety of methods have been devised for the study of protein interactions (Dziembowski and Seraphin 2004). Of these, immunoaffinity purification provides an effective means for isolating protein complexes and elucidating their composition (Kellog and Maozed 2002). Immunoisolation is achieved with antibodies directed either specifically against the proteins of interest or against tags that are coupled to the proteins of interest. The second strategy, utilizing tagged proteins, is attractive because a single highly optimized antibody can serve as the immunoaffinity reagent. Commonly used tags include the FLAG and MYC peptides and Protein A (Einhauer and Jungbauer 2001).

This approach uses immunoaffinity purification on magnetic beads coated with antibodies for the rapid and efficient purification of protein complexes from cells or tissues. This method was originally developed using a green fluorescent protein (GFP) tag for the consecutive visualization and isolation of protein complexes in living systems (Cristea et al. 2005). This methodology has been used successfully to look for interactions in both yeast and mammalian systems, such as studies of the dynamic virus–host protein interactions during the course of a viral infection (Cristea et al. 2006). For teaching the protocol at this course, we use an *S. cerevisiae* strain with the nuclear pore gene *Nup84* fused to the ProteinA tag.

PROTOCOL 5

Conjugation of Magnetic Beads

MATERIALS

CAUTION: See Appendix 11 for appropriate handling of materials marked with <!>.

Reagents

3 M Ammonium sulfate
Anti-GFP antibodies
100 mM Glycine-HCl (pH 2.5)
IgG or high-affinity purified antibodies
PBS
PBS, 0.02% NaN$_3$ <!>
PBS, 0.5% Triton X-100 <!>
0.1 M Sodium phosphate buffer (pH 7.4)
100 mM Triethylamine <!>
  Prepare the triethylamine solution fresh just before use in Step 6.
10 mM Tris-HCl (pH 8.8)

Equipment

Dynabeads, M-270 Epoxy (Dynal/Invitrogen)
Magnetic particle concentrator, 1.5-mL tube type (Dynal/Invitrogen 12020D)
Microcentrifuge tubes, round-bottomed
Neodymium magnets (Dynal/Invitrogen or National Imports MAGCRAFT)

Rare-earth neodymium magnets are used to capture the Dynabeads. Dynal/Invitrogen offers different size magnetic particle concentrators. In addition, National Imports offers a large variety of economical MAGCRAFT neodymium magnets in different shapes and sizes that can be used to capture Dynabeads. The magnets can be temporarily attached to the sides of tubes using rubber bands.

Rotor in a 30°C environment
Tube shaker (e.g., MT-360 Microtube Mixer; Tomy)

PROCEDURE

It is best to carry out Steps 1–5 in the afternoon (~4:00 PM), and wash the conjugated beads the next morning.

1. Weigh the needed amount of magnetic beads in a round-bottom microcentrifuge tube.

   Isolations using 1 mg of beads are good for small-scale pilot experiments. 4 mg of beads is usually sufficient for performing a single isolation, but this amount will depend on the abundance of the protein of interest. 10–20 mg of beads are appropriate when isolating highly abundant proteins.

2. Wash the beads with 1 mL of 0.1 M sodium phosphate buffer (pH 7.4), vortex for 30 seconds, and mix for 15 minutes on a Tomy shaker at room temperature.

3. Place the tube with the bead slurry on the magnet, remove the buffer, and wash the beads again with 1 mL of 0.1 M sodium phosphate buffer (pH 7.4); vortex for 30 seconds, place the tube on the magnet, and remove the buffer.

4. Resuspend the beads with IgG, anti-GFP antibodies, or other antibodies that are to be used for the affinity isolations. Use approximately 20 µL total volume per mg of beads.

   Carefully calculate the amount of antibody and solution volumes to use for each experiment. Use 10 µg of Ab/mg beads for IgG and commercially available antibodies, and 5 µg of Ab/mg beads for purified, high-affinity, custom-made antibodies. Saturation of 1 mg of M-270 beads is achieved with ~7–8 µg of antibody or IgG (Fig. 10). Using more than these specified amounts will lead to an unacceptable background from unbound antibody.

   Here we give an example in which antibodies are conjugated to 10 mg of beads. Prepare the reaction mix as per the table. It is important to add the components in the order given in the first column. Add the 3 M ammonium sulfate last to a final concentration of 1 M.

Reagent	Volume	*Example: To conjugate IgG to 10 mg of beads, then total reaction volume is ~200 µL*
Magnetic beads		10 mg
Antibody solution	V_{Ab} = volume to achieve the desired Ab concentration (see notes above)	e.g., volume to achieve final IgG amount of 100 µg
0.1 M Sodium phosphate buffer (pH 7.4)	$V_T - V_{Ab} - V_{sulf}$	200 µL – V_{Ab} – 66.67 µl
3 M Ammonium sulfate	V_{sulf} = 33% of V_T for a final concentration of 1 M	66.67 µL

V_T = total volume; V_{Ab} = antibody solution volume; V_{sulf} = 3 M ammonium sulfate volume.

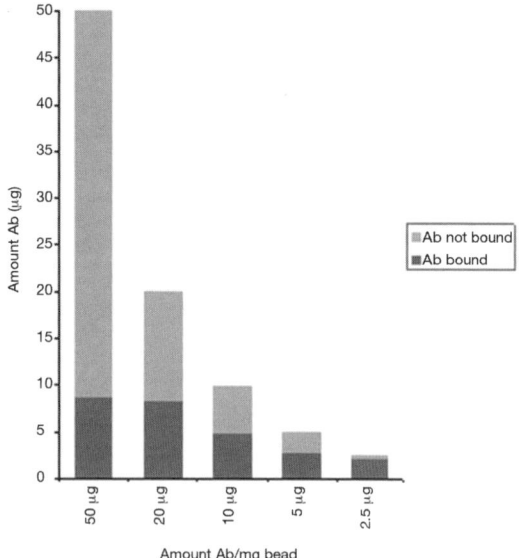

FIGURE 10. Binding capacity of magnetic affinity capture beads for antibodies. The graph shows that maximum binding is achieved with 7–8 µg of antibodies per mg of magnetic beads. (Reprinted, with permission, from the American Society for Biochemistry and Molecular Biology, Inc.)

5. Conjugate the antibodies to the beads overnight on a rotating wheel at 30°C.

6. In the morning, place the tube on a magnet. Remove the supernatant and wash the beads sequentially with:
   - 1 mL of 0.1 M sodium phosphate buffer (pH 7.4)
   - 1 mL of 100 mM glycine·HCl (fast wash)
   - 1 mL of 10 mM Tris (pH 8.8)
   - 1 mL of 100 mM triethylamine (freshly prepared; fast wash)
   - Four washes of 1 mL each with PBS
   - 1 mL of PBS containing 0.5% Triton X-100 for 15 minutes
   - 1 mL of PBS.

   Be careful not to lose magnetic beads during the washes. After each washing step, the washing solution should have a clear aspect, with no trace of beads (Fig. 11).

7. Store the beads at 4°C in PBS, 0.02% NaN_3.

   Beads should be used within 2–3 weeks of conjugation. After 1 month of storage, their isolation efficiency decreases by approximately 40%.

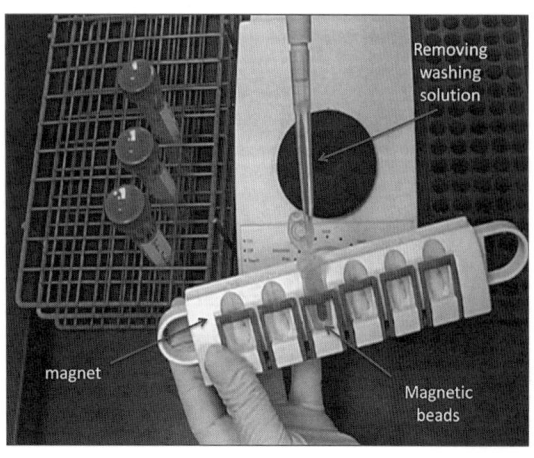

FIGURE 11. Using a magnetic concentrator to capture the magnetic beads for washing. The magnetic beads are held against the side of the tube by the magnetic particle concentrator while the liquid is carefully removed using a clean pipette tip.

PROTOCOL 6

Affinity Purification of Protein Complexes

MATERIALS

CAUTION: See Appendix 11 for appropriate handling of materials marked with <!>.

Reagents

Cell powder, frozen (see Fig. 12)
Lysis buffer, optimized (see Step 1)
Magnetic beads conjugated with the antibodies to be used for the affinity isolation (prepared in Protocol 5)
0.5 N NH_4OH <!>, 0.5 mM EDTA
   Prepare fresh just before use.
SDS-PAGE sample buffer
Stain, mass spectrometry-compatible (for staining proteins in SDS-PAGE gels; see Step 15)

Equipment

Centrifuge (3000 rpm) at 4°C
Liquid nitrogen <!>
Polytron
Rotor chilled to 4°C
Round-bottom tubes for affinity isolations (e.g., culture tubes for small volumes or 50-mL Falcon tubes for larger volumes)
Magnetic particle concentrator, 1.5-mL tube type (Dynal/Invitrogen 12020D)
Microcentrifuge tubes
Neodymium magnets (Dynal/Invitrogen or National Imports MAGCRAFT)
   Rare-earth neodymium magnets are used to capture the Dynabeads. Dynal/Invitrogen offers different size magnetic particle concentrators. In addition, National Imports offers a large variety of economical MAGCRAFT neodymium magnets in different shapes and sizes that can be used to capture Dynabeads. The magnets can be temporarily attached to the sides of tubes using rubber bands.
SDS-PAGE equipment for 1D gels
Shaking incubator set at 70°C
Tomy shaker
Vacuum evaporator (e.g., SpeedVac)

PROCEDURE

1. Prepare the lysis buffer. The buffer should be optimized to give efficient extraction of the tagged protein while being mild enough to maintain the sought after protein–protein interactions.

   During the affinity purification step, the yield of the isolated tagged protein (and any associated macromolecules) is highly dependent on the conditions used for extraction and purification.

Optimize the extraction and purification conditions individually for each different protein complex of interest. Examples of detergents to test are Triton, Tween, deoxycholate, octylglucoside, and digitonin. Various concentrations of salt should also be tested. Examples of buffers are described in Cristea et al. (2005). Begin by testing several buffers of various stringencies. This provides a good indication for what it takes to isolate the tagged protein without losing interacting partners and without accumulating nonspecific interactions. While trying only one buffer is not recommended, a possible starting lysis buffer is 20 mM HEPES-KOH (pH 7.4), 110 mM potassium phosphate, 2 mM $MgCl_2$, 0.1% Tween-20, 0.1% Triton, 150 mM NaCl, 1/100 (v/v) protease inhibitor cocktail. There is no guarantee that this would work on a specific protein complex.

2. Resuspend the frozen cell powder in the optimized lysis buffer (Fig. 12). Use 5 mL buffer per gram of cells and gently mix to ensure resuspension. Do not let the cell pellet thaw before adding the buffer containing the protease inhibitors.

   If the sample derives from cells that were cryogenically ground inside round-bottom microcentrifuge tubes, then open the tubes in a fume hood, add the buffer in small aliquots (e.g., 200 μL), and shake the tube. In this way, the stainless steel grinding ball is used to wash the tube several times.

3. To improve the extraction, homogenize the cells for 10–15 seconds using a Polytron.

   Keep the cell lysate on ice. This step will generate foam. It is advisable to use a container in which the cell lysate will occupy less than 1/3 of the volume during the Polytron step. This will ensure that sample will not be lost during the procedure.

4. If necessary, place the cell lysate on a rotor (gentle rotation) for 5–10 minutes at 4°C to reduce the amount of foam.

5. Centrifuge the cell lysate at 3000 rpm for 10 minutes at 4°C.

   Any remaining foam will disappear during the centrifugation.

6. While the cell lysate is being centrifuged, measure the amount of conjugated beads needed to perform the immunoprecipitation. Place them on a magnet, remove the PBS, 0.02% NaN_3 solution, and wash the beads three times each with 1 mL of lysis buffer. After the third wash, resuspend the beads in a small volume of lysis buffer (e.g., 50–100 μL).

7. Transfer cleared cell lysate to a clean container. Preferably use a round-bottom or wide-bottom tube (e.g., culture tubes for smaller volumes or 50-mL Falcon tubes for larger volumes).

8. Check the cleared cell lysate carefully for any particles that might not have pelleted.

   The supernatant should be clear. If any particles are present, remove them, as these can bind/block the beads and interfere with the isolation efficiency or cause a large unwanted background. Additional centrifugation may not pellet the particles (e.g., lipids) depending on the cells or tissue. In these cases, the particles can be removed using a pipetman.

9. Add the washed beads (from Step 6) to the cell lysate. Incubate at 4°C with gentle rotation for 5 minutes to 1 hour.

   Ensure that beads are in contact with the cell lysate and do not get trapped on the walls or cap of the container. Do not use longer incubation times, as this promotes the accumulation of nonspecific binding and the loss of weak interacting partners (Fig. 13). The optimal incubation time must be determined empirically.

10. After incubation, place the tube on a magnet. Transfer the flowthrough to another tube to use for Western blot analyses to measure the efficiency of protein recovery.

11. Wash the beads six times with 1 mL of lysis buffer. During the first wash, transfer the beads to a clean microcentrifuge tube. After the fourth wash, transfer the beads to another clean microcentrifuge tube. This will ensure that any components of the cell lysate that might have bound to the walls of the tube will not contaminate the eluate.

This protocol uses cells that have been cryogenically disrupted with a ball mill (e.g., Retsch MM301) or a mortar grinder as a starting sample (Fig. 12). However, the same protocol can be used for cells that were disrupted by incubation with lysis buffers, glass beads, or passage through needles of various gauges. We prefer the cryogenic cell disruption because it significantly increases the extraction efficiency and it has proven to be absolutely critical in most of our studies (Cristea 2005, 2006; Wang 2006). For studies that require gentle handling of the cells in order to isolate large intact organelles, the other mentioned techniques might yield better results.

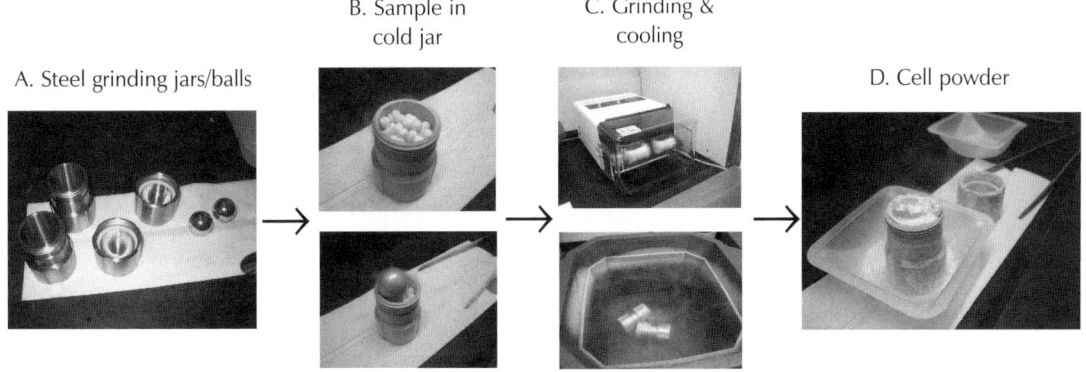

FIGURE 12. Experimental flow chart for preparing cell extracts by cryogenic grinding of cells or tissues. (A) Stainless steel jars, lids, and balls for cryogenically disrupting frozen cells. The stainless steel jars and balls are used to cryogenically grind frozen cells into a powder for making whole cell protein extracts. Before use, the jars and balls are precooled in liquid nitrogen. (B) Frozen cell pellets added to a frozen steel jar. The pellets are made by dropping a cell slurry into liquid nitrogen and recovering the frozen cell pellets. The precooled steel ball is added to the jar with the cell pellets. The stainless steel cap is then screwed on to seal the chamber. (C) Steel jars inserted into the grinding mill (Retsch MM 301 Mixer Mill). For yeast, the cells are ground into a powder using 10x 3 minute cycles at 25 Hz. Between each grinding cycle, the jars are removed from the holders and plunged into liquid nitrogen to keep the cells frozen. (D) Cryogenically disrupted cells. A fine powder is created from the frozen cell pellets and grinding process. The powder is kept frozen and quickly transferred to a 50 mL tube kept cool in liquid nitrogen using a cold spatula. The disrupted cells are stored at –80°C until ready to use.

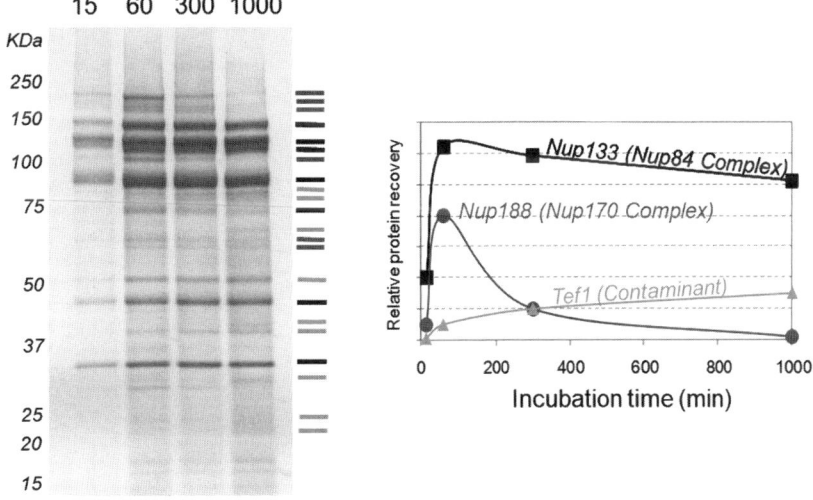

FIGURE 13. Effects of incubation times on protein recovery and nonspecific interactions. Coomassie-stained SDS-PAGE showing isolation of GFP-tagged Nup84 and its associated proteins using various incubation times (*left*). Recoveries of the indicated proteins as a function of time. Proteins were identified by mass spectrometry (*right*). (Reprinted, with permission, from the American Society for Biochemistry and Molecular Biology, Inc.)

12. After the sixth wash, elute the isolated proteins from the beads.

    a. Add 500 μL of freshly made aqueous 0.5 N NH$_4$OH, 0.5 mM EDTA solution to the beads.

    b. Shake the tube (e.g., using a Tomy shaker) for 20 minutes at room temperature.

    Place the tube on a magnet and transfer the eluate to a clean microcentrifuge tube. Keep the beads to test for the efficiency of elution.

    > Use a larger volume when the amount of beads used for the isolation exceeds 10 mg. The elution can also be performed with other solutions. For example, acid elutions can be achieved with 0.1 M citrate (pH 3.1; as recommended by Dynal), or 0.1% TFA (pH 1.5). In our hands, the elution with citrate is not very efficient.

13. Snap-freeze the eluate in liquid nitrogen and dry it in a vacuum evaporator overnight (a minimum of 4 hours). Ensure that the sample is fully dried before proceeding.

14. Add SDS sample buffer (e.g., 20 μL) to the dried sample. Shake on a Tomy shaker for 10 minutes, and place the sample at 70°C (if possible with shaking) for 10 minutes. Keep 10% (e.g., 2 μL) of the sample for Western blot analysis to test for IP efficiency.

15. Run the sample on a one-dimensional SDS-PAGE gel. Stain the gel with a stain that is compatible with mass spectrometric analysis (e.g., colloidal Coomassie Blue or zinc staining).

    > To identify the purified proteins, individual bands can be in-gel trypsin digested (Appendix 3) and the recovered peptides analyzed by tandem mass spectrometry (Experiment 3 or 4). Alternatively, the entire gel lane of the sample can be cut into equal pieces, each gel piece in-gel trypsin digested, and recovered peptides analyzed by tandem mass spectrometry. Because of the high sensitivity of mass spectrometry, it is common to detect proteins in unstained regions of the gel that appear to be void of proteins.

REFERENCES

Cristea I.M., Williams R., Chait B.T., and Rout M.P. 2005. Fluorescent proteins as proteomic probes. *Mol. Cell. Proteomics* **4**: 1933–1941.

Cristea I.M., Carroll J.W., Rout M.P., Rice C.M., Chait B.T., and MacDonald M.R. 2006. Tracking and elucidating alphavirus-host protein interactions. *J. Biol. Chem.* **281**: 30269–30278.

Dziembowski A. and Seraphin B. 2004. Recent developments in the analysis of protein complexes. *FEBS Lett.* **556**: 1–6.

Einhauer A. and Jungbauer A. 2001. The FLAG peptide, a versatile fusion tag for the purification of recombinant proteins. *J. Biochem. Biophys. Methods* **49**: 455–465.

Gavin A.C., Aloy P., Grandi P., Krause R., Boesche M., Marzioch M., Rau C., Jensen L.J., Bastuck S., Dümpelfeld B., et al. 2006. Proteome survey reveals modularity of the yeast cell machinery. *Nature* **440**: 631–636. Epub 2006 Jan 22.

Ghaemmaghami S., Huh W.K., Bower K., Howson R.W., Belle A., Dephoure N., O'Shea E.K., and Weissman J.S. 2003. Global analysis of protein expression in yeast. *Nature* **425**: 737–741.

Kellogg D.R. and Moazed D. 2002. Protein- and immunoaffinity purification of multiprotein complexes. *Methods Enzymol.* **351**: 172–183.

Rigaut G., Shevchenko A., Rutz B., Wilm M., Mann M., and Séraphin B. 1999. A generic protein purification method for protein complex characterization and proteome exploration. *Nat. Biotechnol.* **17**:1030–1032.

Wang Q.J., Ding Y., Kohtz D.S., Mizushima N., Cristea I.M., Rout M.P., Chait B.T., Zhong Y., Heintz N., and Yue Z. 2006. Induction of autophagy in axonal dystrophy and degeneration. *J. Neurosci.* **26**: 8057–8068.

EXPERIMENT 3

Qualitative and Quantitative Measurement of Peptides with MALDI TOF/TOF Mass Spectrometry

Eric S. Simon

*Department of Biological Chemistry, University of Michigan,
Ann Arbor, Michigan 48109*

OVERVIEW OF MALDI TOF/TOF MASS SPECTROMETRY

To analyze samples by mass spectrometry, the analyte must be in the gas phase and must have a net charge. Matrix-assisted laser desorption ionization (MALDI) was introduced in the 1980s as a "soft ionization technique" capable of analyzing large molecules that tend be fragile and fragment when ionized by other methods (Karas and Hillenkamp 1988). MALDI is an efficient process for generating gas-phase ions of peptides and proteins for mass spectrometric detection. In MALDI, a solution containing the analyte is codeposited on a clean surface, or MALDI target plate, with a solution containing a high concentration of a UV-absorbing molecule, known as a matrix (Karas and Hillenkamp 1988). The deposited sample and matrix solution dry and crystallize. For peptides, the predominantly used matrices are 4-hydroxy-α-cyanocinnamic acid (HCCA) and 2,5-dihydroxy benzoic acid (DHB). The plate with the crystallized droplet(s) is injected into the ionization chamber of a mass spectrometer, which is pumped down to a pressure of approximately 10^{-7} Torr. In a process that is still incompletely understood, analyte and matrix molecules are ablated and ionized by protonation in the gas phase by a series of pulsed laser shots on the sample spot (Karas et al. 2000). The crystallized matrix molecules on the sample spot absorb the bulk of the irradiation, but enough energy is transferred to the analyte molecules to induce desorption from the surface and ionization. MALDI results in intact molecular ions whose net charges are almost exclusively +1.

Once the analyte molecules have been desorbed and ionized, they are typically accelerated down a flight tube under the influence of an electric field toward a detector. This is a mass analyzer known as time-of-flight (TOF) in which the flight time of the ion from the source to the detector, t, is correlated to the *m/z* of the ion (Wolff and Stephens 1953). In general, larger mass ions take more time to travel the length of the flight tube than smaller mass ions. A TOF mass analyzer typically operates in one of two modes. In the linear mode, ions accelerate from the source toward a detector at the opposite end of the flight tube (Wolff and Stephens 1953). In the reflector mode, the ions traverse the flight tube and, before they reach the linear detector, are deflected back down the opposite direction of the flight tube to another detector (Mamyrin et al. 1973). This results in significantly improved resolution and mass spectra as compared to spectra acquired in the linear mode.

A MALDI TOF/TOF mass spectrometer is a MALDI TOF instrument with the added capability for efficiently generating MS/MS spectra by fragmenting selected precursor ions desorbed from the sample plate by MALDI and measuring the fragments via a TOF analyzer. A misconception, implied by the name TOF/TOF, is that this instrument is composed of two flight tubes arranged in tandem. On the contrary, it consists of just one flight tube. Part of the flight path includes an evacuated chamber that ions simply pass through on their way to the detector when the instrument is operated in MS mode. However, when operated in MS/MS mode, the chamber is filled with atmospheric gas that interacts with and induces fragmentation of ions entering the chamber. The first TOF term in TOF/TOF, which will be called TOF1 for simplicity, represents the time required for an ion of interest, with a specific m/z value, to reach the gas-filled collision chamber. TOF1 is a critical parameter for performing MS/MS on a MALDI TOF/TOF mass spectrometer. The ions leaving the MALDI source and accelerating toward the collision chamber have a wide range of masses, but it would be ideal to have only the ion of interest enter the chamber. This is addressed by a device called a timed ion selector (TIS). It deflects all ions away from the entrance to the collision chamber until the precise time (TOF1) that the ion of interest arrives at the chamber. The deflectors are then turned off, allowing the selected ion to enter the collision cell and fragment. The deflectors are then turned back on to prevent higher mass ions from entering the collision chamber. The fragmented ions generated in the collision chamber are accelerated toward a detector (in either a linear or reflector flight path) where their time of flight (TOF2) is measured and recorded. Hence, the term TOF/TOF refers to the time required for a selected precursor ion of interest to reach a collision chamber (TOF1) and the time for the resulting fragments to reach the detector (TOF2).

PROTOCOL 1

Experimental Description

The following procedures highlight, in a generalized manner, the steps for setting up the 4700 MALDI TOF/TOF mass spectrometer and acquiring data from samples generated from the protocols in Experiments 1 (2D gels) and 7 (iTRAQ). Among the steps included are spotting calibration mixture on the MALDI target plates, setting up spot sets for each plate, loading the plate into the mass spectrometer, calibrating the plate, and finally, acquiring data from the plate. Key terms and concepts are defined below. These should be read and referred to during the experiment.

KEY TERMS AND CONCEPTS

- **4000 Series Explorer:** The control software for the 4700 MALDI TOF/TOF mass spectrometer. Figure 1 displays the main window illustrating the essential features used for the proteomics course, including the toolbar, spot set window, and spectrum viewer.

- **Acquisition method:** Specifies the instrument settings needed to acquire data on the 4700 MALDI TOF/TOF mass spectrometer. It specifies the mass range, focus mass (mass of maximum resolution), number of laser shots per spectrum, laser intensity, laser firing pattern within each well, and operating mode. The operating mode specifies the mode of acquisition (positive or negative ions), whether it is an MS method (precursor measurement) or MS/MS method (collision-induced dissociation of a selected precursor ion), and whether to use detection of ions from the linear detector or reflector detector.

- **Batch mode acquisition:** Automatic acquisition of data from a series of preselected spots selected within the spot set.

- **Interactive mode acquisition:** Manual acquisition of data from a sample spot.

- **Interpretation method:** Specifies parameters for selecting precursor ions for MS/MS acquisition by interpreting the previously acquired MS spectrum or spectra. It specifies a minimum

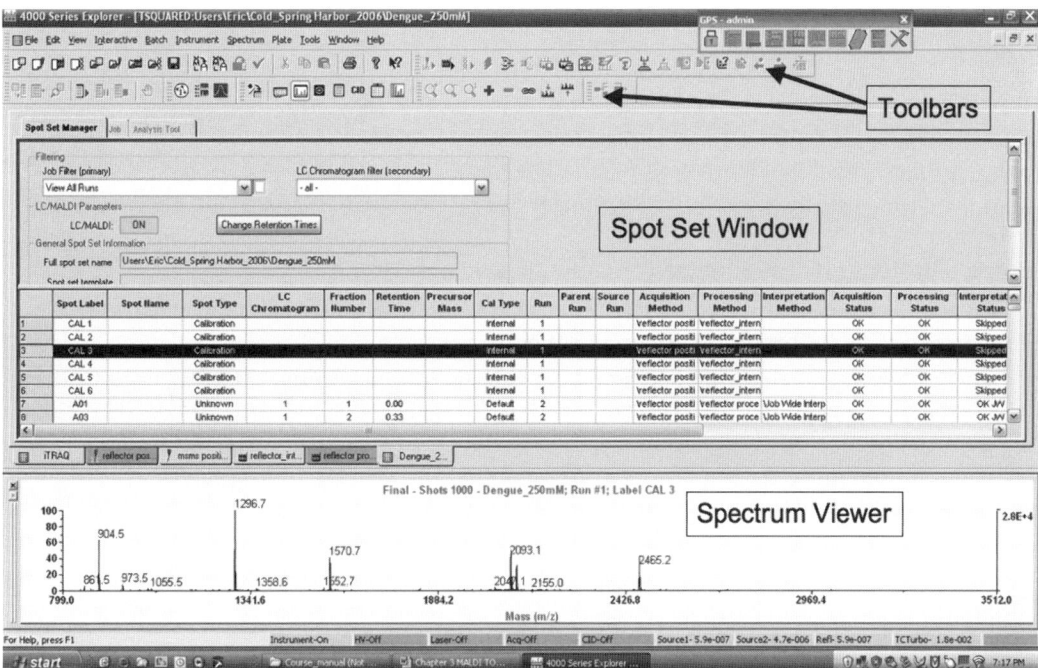

FIGURE 1. Main window and general view of the 4000 Series Explorer operating software for the 4700 MALDI TOF/TOF mass spectrometer.

SN threshold to select a precursor for MS/MS, an order of acquisition (e.g., acquire MS/MS spectra of the top eight most intense precursors from each spot), and the MS/MS acquisition and processing methods to use. The user can also input peaks for inclusion (acquire MS/MS if present) or exclusion (ignore if present).

- **Job:** A list of spots queued for acquisition that includes the acquisition, processing, and interpretation method (if necessary) to be used.
- **Job-wide interpretation method:** Used for LC-MS applications to eliminate redundant precursor MS/MS acquisitions. For example, if a precursor ion is detected in three consecutive spots, it will only acquire from the spot where the precursor ion was most intense.
- **Oracle:** All methods and data are stored in an Oracle database.
- **Processing method:** Contains parameters for handling data once it has been acquired. It includes options for smoothing data, settings for peak detection like signal-to-noise, SN, thresholds, and calibration (mass determination).
- **Project:** This is essentially equivalent to a data directory. For the proteomics course, the project is called "proteomicscourse," which holds all relevant spot sets and methods as they are created during the course.
- **Sample plate:** Also called a MALDI target plate. A plate provides a surface, typically stainless steel, with predefined sample wells, or spots, where samples have been codeposited with a matrix that crystallizes after drying. Each plate has a unique name stored in the database. Also included in the database are the barcode, calibration data, spot template, and alignment for each sample plate. For the course, the sample plates that are used contain 192 sample wells and 6 calibration wells.
- **Spot set:** Contains information, stored in Oracle, on a specific set of samples spotted onto the sample plate. The name of the spot set often reflects the sample and experimental preparation involved.
- **Spot set template:** Specifies the number and layout of spots on the plate.

58 EXPERIMENT 3

MATERIALS

CAUTION: See Appendix 11 for appropriate handling of materials marked with <!>.

Reagents

Matrix solution (α-Cyano-4-hydroxycinnamic acid <!>; HCCA)
  Prepare 5 mg/mL of HCCA in 50% acetonitrile <!> and 0.1% trifluoroacetic acid <!>.
Standard peptide calibration mixture (Applied Biosystems, Inc. 4333604)
  The peptides and their corresponding masses are des-arg1-bradykinin (904.5 Da), angiotensin I (1296.7 Da), glu1-fibrinopeptide B (1570.7 Da), ACTH (1–17) (2093.1 Da), and ACTH (18–39) (2465.2 Da). The final concentration of each peptide is 1 pmole/μL.
Trypsin-digested protein samples previously spotted on MALDI target plates (refer to Experiments 1 and 7)

Equipment

4000 Series Explorer software for operating the 4700 MALDI TOF/TOF mass spectrometer (Applied Biosystems, Inc.)
MALDI target plate(s), 192-well stainless steel (Applied Biosystems, Inc. V700666)
MALDI TOF/TOF mass spectrometer (model 4700; Applied Biosystems, Inc.)
Pipette and pipette tips

PROCEDURE

Spotting Calibration Solution onto the Calibration Wells

1. Spot 0.5 μL of calibration mixture on calibration well 1 (CAL1).
2. Before the solution dries, spot 0.5 μL of matrix solution on CAL1.
3. Repeat for each of the five remaining calibration wells.

Creating a New Spot Set

Before loading a new sample plate into the 4700 MALDI TOF/TOF mass spectrometer, create a new "spot set," a new "plate" associated with the spot set, and choose a "spot set template" for the spot set. Each MALDI plate loaded into the 4700 Proteomics Analyzer must have a unique spot set associated with it. If you wash and respot a plate, you must create a new spot set for the plate.

1. Select "File" > "New" > "Spot Set." The "Create New Spot Set" dialog box opens.
2. From the "Project" drop-down list, select the project name (in the course, choose "proteomicscourse").
3. Name the spot set. In the "Item Name" field (at the bottom of the dialog box), type in the name of the sample plate. Although any name can be used, it is convenient to use the plate's barcode as a name.

QUALITATIVE AND QUANTITATIVE MEASURE OF PEPTIDES 59

4. Enter the plate barcode into the database.

   In the proteomics course, all of the plate barcodes have been stored in advance, so you will not need to enter that information.

5. Click "Create."

6. Name the plate. In the "Select/Create Plate for New Spot Set" window, type in a plate name in the "Plate Name" field.

   In the course, name the plate by group name and experiment. For example, "Group A 2D gels" or "Group C iTRAQ IEF3."

7. Click "OK."

8. Select "Spot Set Template." In the "Select Spot Set Template for New Spot Set" window, a drop-down list appears.

   For 192-well plates spotted with 2D gel samples:

   a. Select "Factory Spot Set Template." A number of factory default templates appear with descriptive columns.

   b. Select "ABI-192+6AB." This template is named after its plate type and represents a 192-well sample plate with 6 calibration wells in two columns (A and B). Under "Type," the template is categorized as "N/A," which essentially means that it is not an LC-MS plate and can be used for 2D gel analysis.

   For 192-well plates spotted with LC-separated fractions (e.g., iTRAQ experiments):

   a. Select "User Defined Spot Set Template." There is not a default template for this application with a 192-well plate. For the course, select the template "MPC Projects\LCMS 192 alphanumeric," which was created in advance.

   b. Select "ABI-192+6AB."

   c. Under "Type," categorize the template as "LCMS."

9. Click "Select."

Loading the Sample Plate

Before loading the sample plate, make sure all sample spots are dry, and apply a stream of compressed, dry gas over the plate to remove fibers that could be introduced into the instrument.

1. Select "Plate" > "Eject Plate." The "4000 Series Explorer Plate Manager" dialog box opens, asking "Are you sure you want to eject plate?" Click "OK." Wait for sample holder to appear.

2. Insert sample plate into plate holder. Hold the sample plate vertically, with the top of the plate up and the sample surface facing to the right. Slide the sample plate into the holder until the ball bearings on the holder snap into the plate. Figure 2 shows the proper orientation of the sample plate in the plate holder.

3. Select "Plate" > "Load Plate." The "Select Spot Set" dialog box opens.

4. Select the appropriate project and spot set. For the proteomics course, from the "Projects" drop-down list, select "proteomicscourse." Select the spot set that corresponds to the sample plate that has been inserted into the plate holder. For example, "Group A 2D gels" or "Group C iTRAQ IEF3."

5. The "Load Sample Plate" dialog box appears, displaying details about the plate and spot set selected for loading. Click "Load."

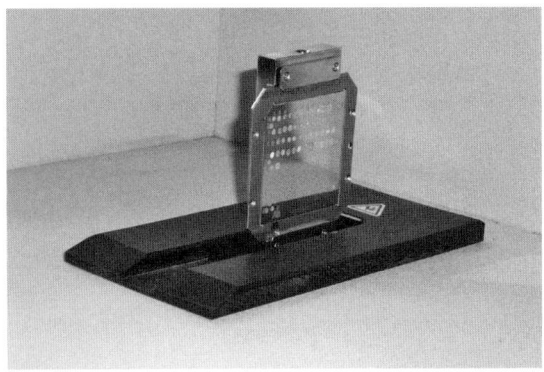

FIGURE 2. The proper orientation of the MALDI sample plate in the plate holder.

Important: Keep hands clear of the plate holder during loading. The plate loads and the source chamber pumps down. A window appears that displays the status of the "Load/Eject Cycle." Once the window disappears, the instrument is ready for acquisition.

6. Once the plate is loaded, a spot set window appears with three tabs that can be selected. By default, the "Spot Set Manager" tab is selected, displaying a spreadsheet (representing the spot set) in the main window above a selection tab for the spot set (Fig. 3).

Calibrating the Sample Plate in Reflector Mode

To acquire data, acquisition and processing methods are activated. Data acquisition for all applications for the proteomics course will be performed in the batch mode. No methods need to be

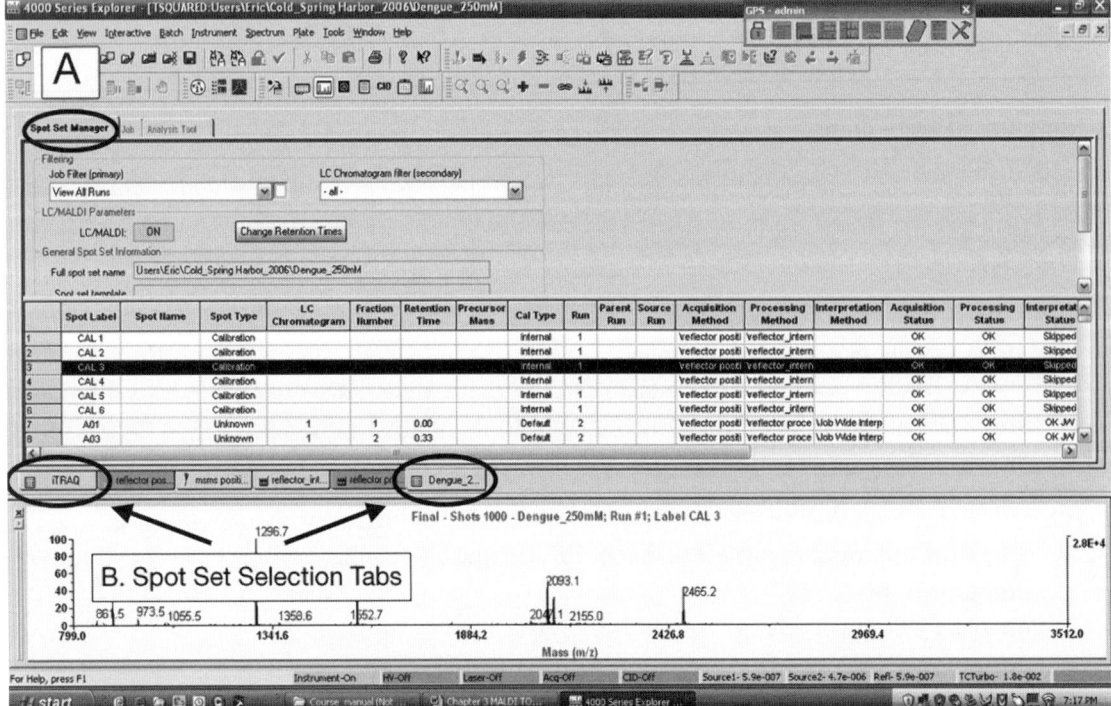

FIGURE 3. The "Spot Set Manager" tab (A) opens by default when a spot set selection tab is selected (B).

opened interactively for this purpose. However, to warm up the electronics (Step 3 below), an acquisition and processing method need to be opened (Steps 1 and 2).

1. Select "File" > "Open" > "Acquisition Method" or click the acquisition icon. From the Project drop-down list, select "MS Reflector Positive." This is a method for acquiring data in MS mode for positive ions using the reflector flight path. An acquisition selection tab below the main window appears, displaying the acquisition method icon and the name of the method (Fig. 4). When the acquisition method selection tab is clicked, the acquisition parameters are displayed in the main window (Fig. 4).

2. Select "File" > "Open" > "Processing Method" or click the processing icon. From the Project drop-down list, select "reflector internal." A processing selection tab appears in the main window, displaying the processing icon and the name of the processing method (Fig. 5). This is an internal processing method that looks for peaks in the spectrum acquired from a calibration well that has been spotted with a calibration mixture composed of five standard peptides (listed in Reagents). The processing method instructs the instrument to look for each of the peptide *m/z* values within a fixed window specified by the method. Once it finds the corresponding peaks for each standard peptide, it adjusts the masses of each appropriately. If data is acquired in the interactive mode (which is the default mode) and there are multiple acquisition and/or processing methods open, the tabs of the active methods are green and the inactive methods are gray (Fig. 5). To activate a different method, click the method selection tab and click.

3. Select "Instrument" > "High Voltage" or click. Allow the high voltage electronics to warm up for 30 minutes before beginning acquisition. If the instrument has been running, this may not be necessary.

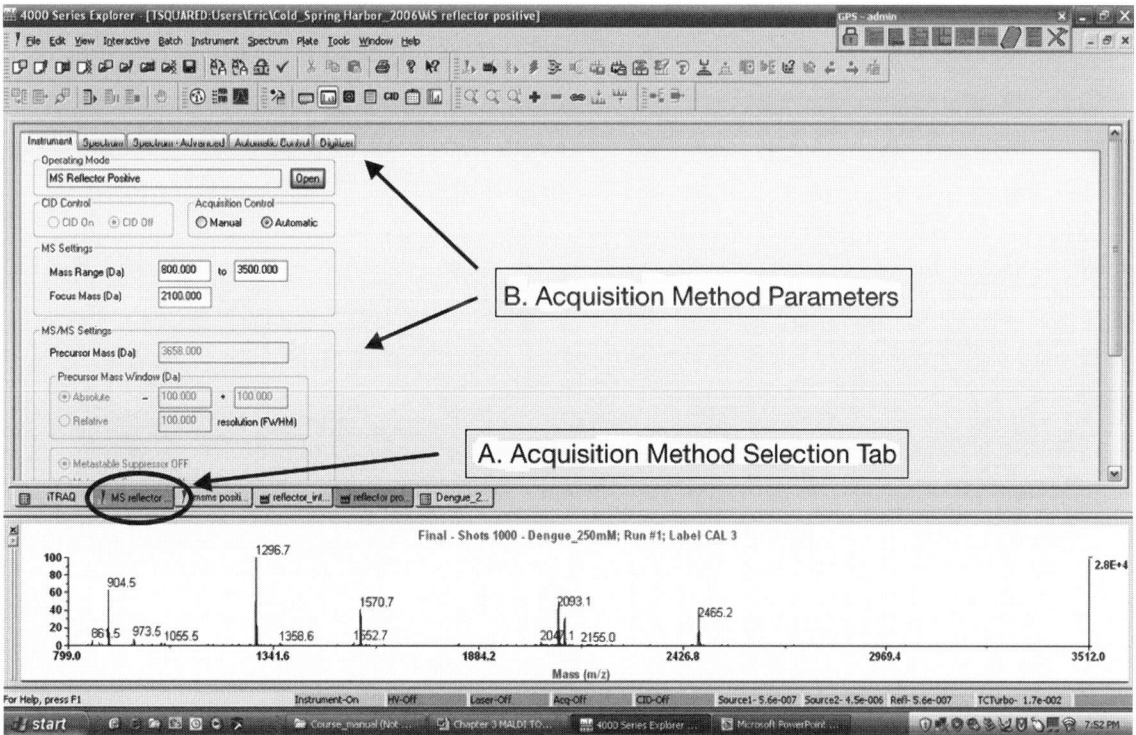

FIGURE 4. Clicking the acquisition method tab (A) opens up the method parameters in the main window (B).

62   EXPERIMENT 3

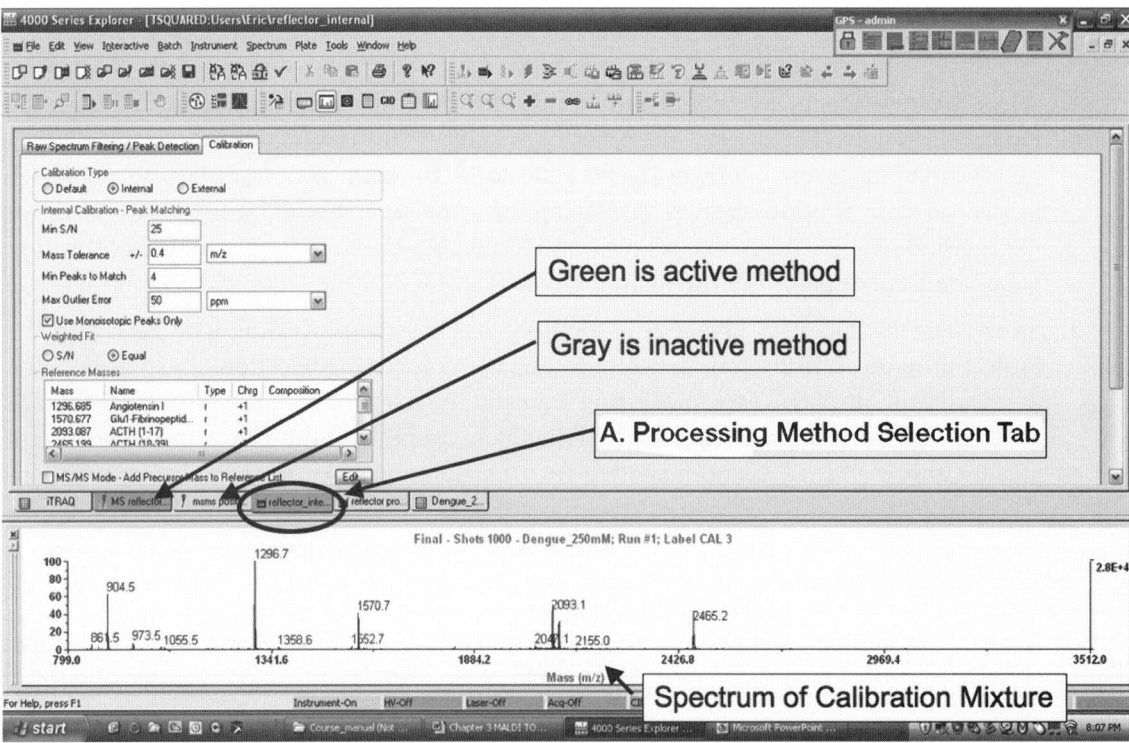

FIGURE 5. Clicking the processing method tab (A) opens up the method parameters in the main window.

4. Select the spot set tab in the main window (Fig. 6).
5. In the spot set window, highlight all of the rows representing the six calibration wells. This is done by clicking each of the numbers in the column to the left of the "Spot Label" column (Fig. 6).

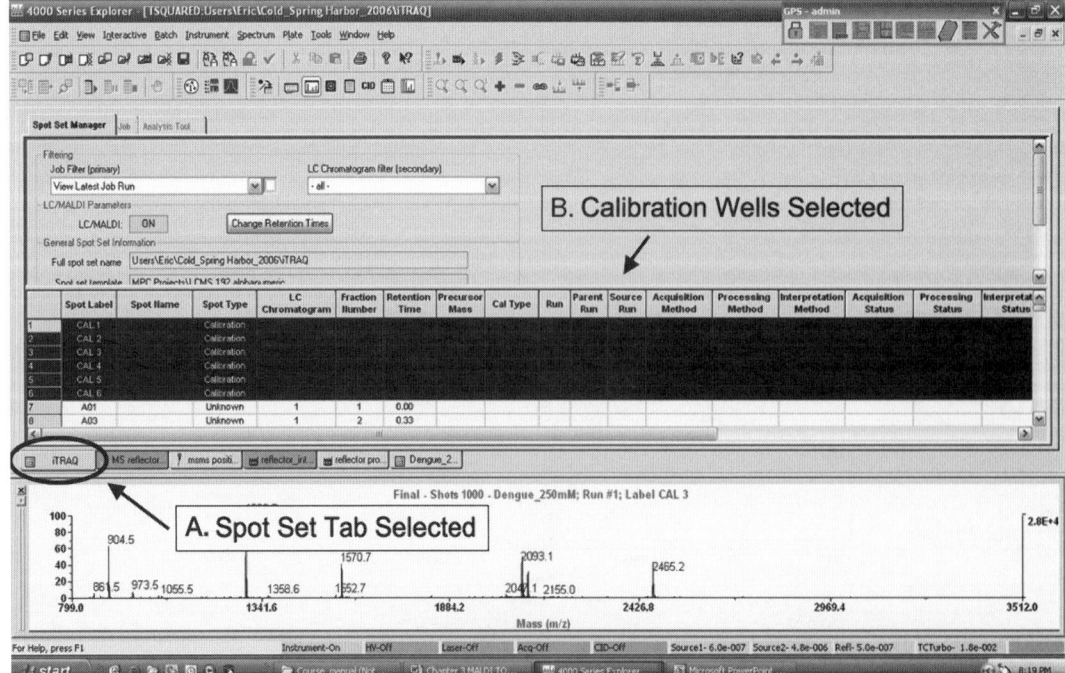

FIGURE 6. Selection of the spot set (A) and selection of calibration rows (B) by clicking the numbers to the left of each spot label.

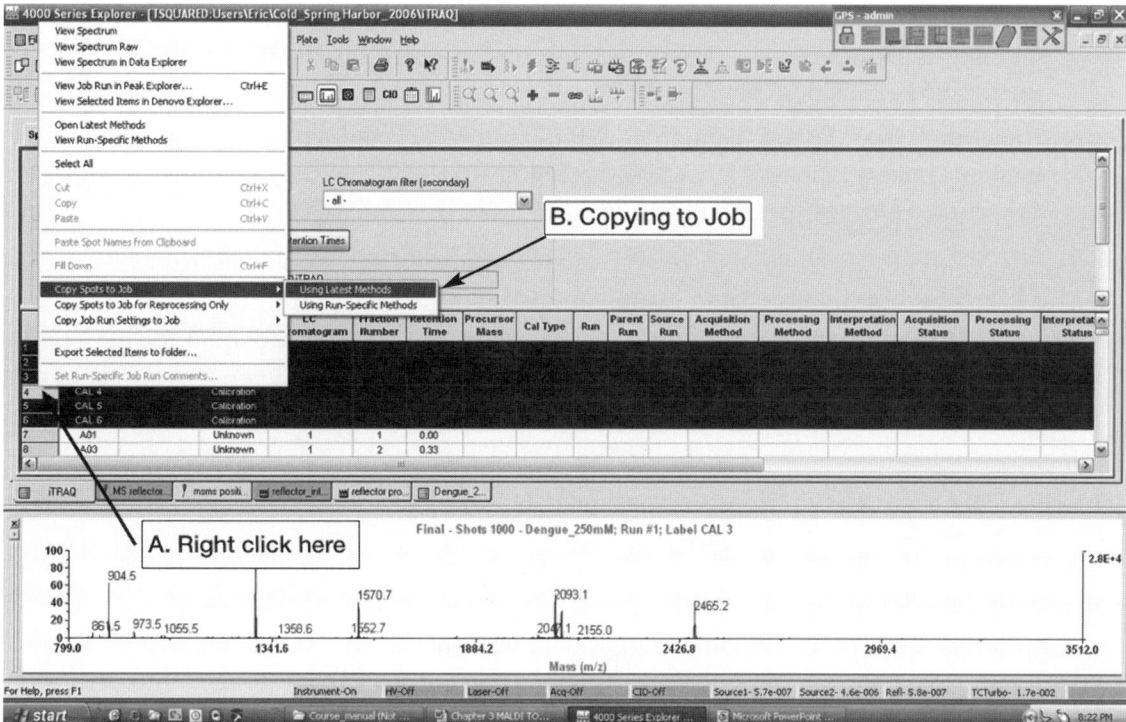

FIGURE 7. Copying the selected rows (or wells) to a job.

6. Right click the highlighted area (Fig. 7) and click "Copy Spots to Job" > "Using Latest Methods" (Fig. 7). In the spot set window, the "Job" tab will now be selected and displays only the rows that have been selected as a job (Fig. 8). Under "Cal Types Updated," select "Plate Model & Default Calibration" (Fig. 8).

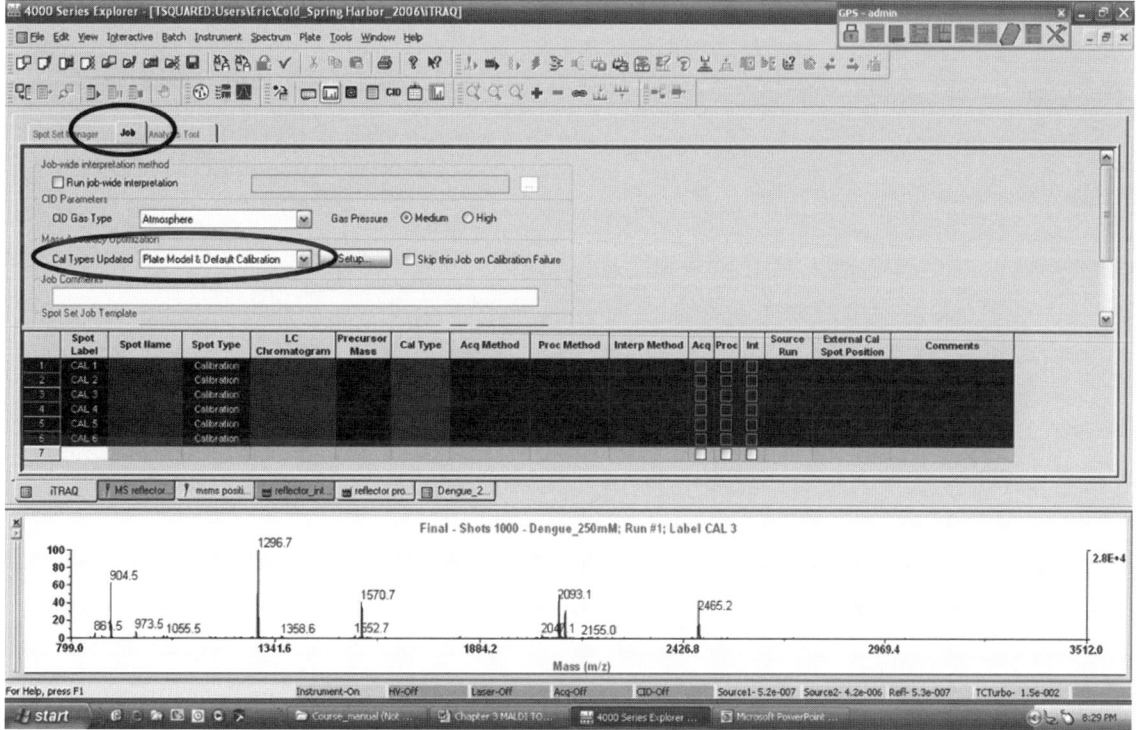

FIGURE 8. The Job window.

64 EXPERIMENT 3

7. Under the "Acq Method" column and in the top cell, click the down arrow and select the method "reflector positive." The software automatically fills in the rest of the column with the selected acquisition method (Fig. 9).

8. Under the "Proc Method" column and in the top cell, click the down arrow and select the method "reflector internal." The software automatically fills in the rest of the column with the selected processing method (Fig. 9).

9. Select "Batch" > "Submit Spot Set Job" or click ▣. This submits the selected job to queue.

10. Click ▶ in the toolbar. The instrument automatically switches to the batch mode and begins acquisition and processing. When finished, the default calibration for the sample plate is updated.

Acquiring Data from 2D Gel Digests (Experiment 1)

1. Calibrate the plate as described in the previous section, Calibrating the Sample Plate in Reflector Mode.

2. In the spot set window, highlight all of the rows representing the sample wells to be analyzed. This is done by clicking each of the numbers in the column to the left of the "Spot Label" column.

3. Right click the highlighted area and click "Copy Spots to Job" > "Using Latest Methods." In the spot set window, the "Job" tab will now be selected and displays only the rows that have been selected as a job. Under "Cal Types Updated," make sure to select "none," since no calibration will be performed in this job (Fig. 10).

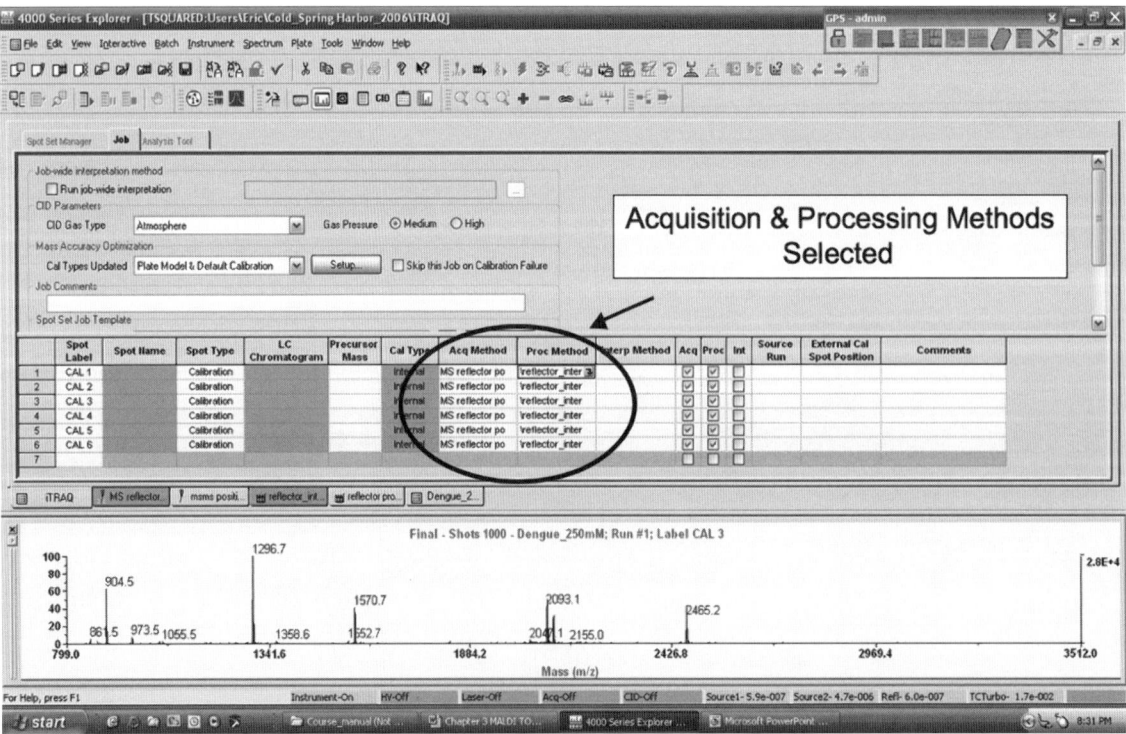

FIGURE 9. Addition of the acquisition and processing methods to the spot set list for the selected calibration job.

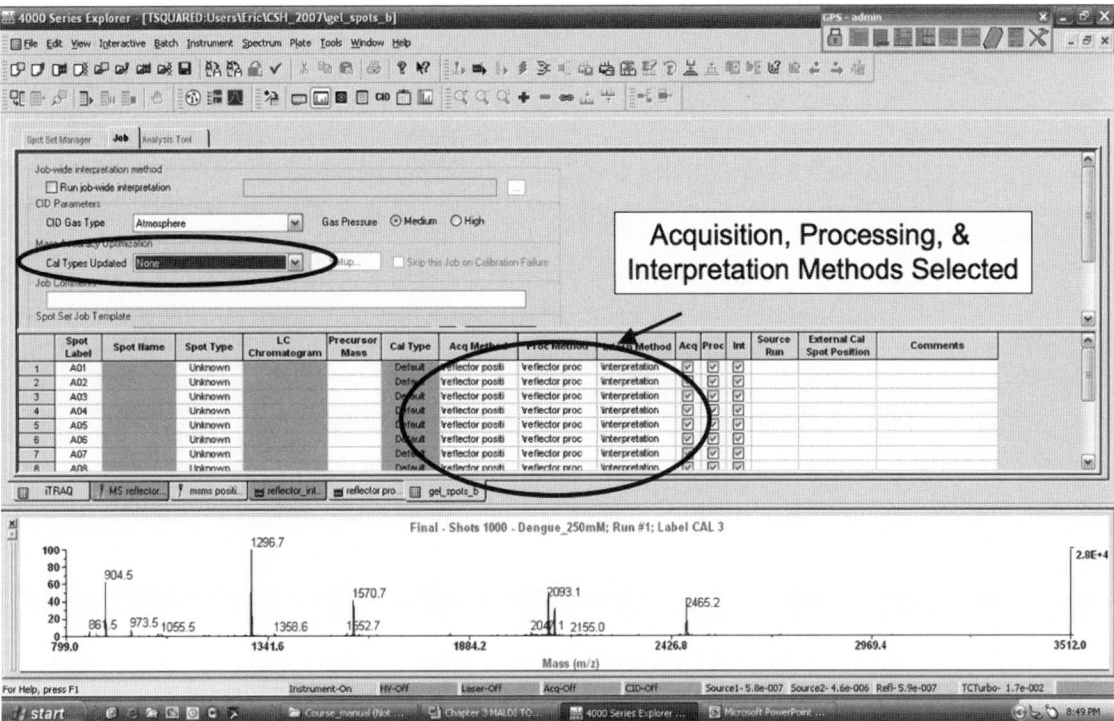

FIGURE 10. Job inputs and settings for analyzing protein digests from 2D-gel slices.

4. Under the "Acq Method" column and in the top cell, click the down arrow and select the method "reflector positive." The software automatically fills in the rest of the column with the selected acquisition method (Fig. 10).

5. Under the "Proc Method" column and in the top cell, click the down arrow and select the method "reflector processing default." The software automatically fills in the rest of the column with the selected processing method (Fig. 10). Since these samples were from in-gel tryptic digestions, this processing method uses tryptic autolysis products as internal standards for mass calibration if they are detected. Otherwise, it uses the default calibration (see previous section) to assign m/z values to peaks.

6. Under the "Interp Method" column and in the top cell, click the down arrow and select "interpretation method." This interpretation selects precursor ions represented by the eight most intense peaks in the MS spectrum from each sample well for MS/MS acquisition. It excludes any detected trypsin autolysis peaks for MS/MS acquisition. The software automatically fills in the rest of the column with the selected interpretation method (Fig. 10).

7. Select "Batch" > "Submit Spot Set Job" or click ▣. This submits the selected job to the queue.

8. Click ▣ in the toolbar. The instrument automatically switches to the batch mode and begins acquisition and processing.

9. Upon completion of the run, process the acquired MS/MS data using either ProteinPilot or Mascot protein database search algorithms to identify the peptides and proteins (refer to Experiment 8).

Acquiring Data from iTRAQ IEF Fractions (Experiment 7)

1. Calibrate the plate as described in the section Calibrating the Sample Plate in Reflector Mode.

2. In the spot set window, highlight all of the rows representing the sample wells to be analyzed. This is done by clicking each of the numbers in the column to the left of the "Spot Label" column.

3. Right click the highlighted area and click "Copy Spots to Job" > "Using Latest Methods." In the spot set window, the "Job" tab will now be selected and displays only the rows that have been selected as a job. Under "Cal Types Updated," make sure to select "none," since no calibration will be performed in this job (Fig. 11).

4. Under the "Acq Method" column and in the top cell, click the down arrow and select the method "reflector positive." The software automatically fills in the rest of the column with the selected acquisition method.

5. Under the "Proc Method" column and in the top cell, click the down arrow and select the method "reflector processing default." The software automatically fills in the rest of the column with the selected processing method. This processing method uses the default calibration (as in the section, Calibrating the Sample Plate in Reflector Mode) to assign m/z values to peaks.

6. Under the "Interpretation Method" column and in the top cell, click the down arrow and select the method "Job Wide Interpretation-LCMS." The software automatically fills in the rest of the column with the selected interpretation method. Since a job-wide interpretation method was selected, the "Run job-wide interpretation" check box is checked (Fig. 11).

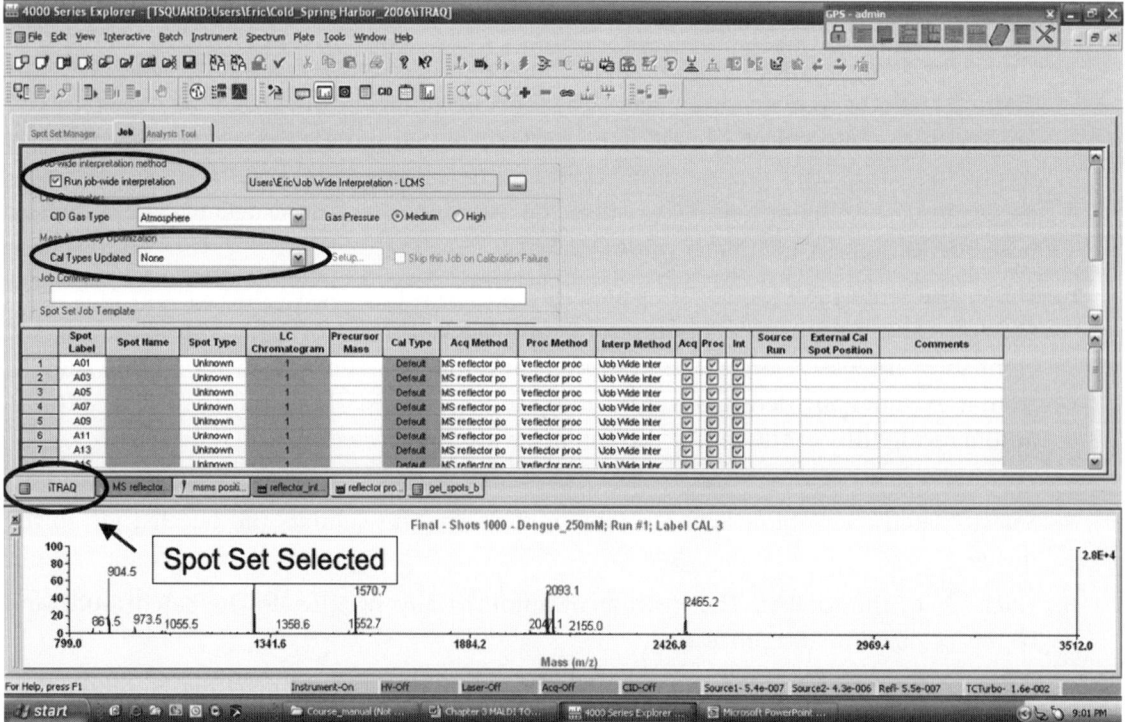

FIGURE 11. Job inputs and settings for analyzing iTRAQ-labeled peptides from IEF fractions.

7. Select "Batch" > "Submit Spot Set Job" or click ▣. This submits the selected job to queue.

8. Click ▣ in the toolbar. The instrument automatically switches to the batch mode and begins acquisition and processing.

9. Upon completion of the run, process the acquired MS/MS data using either ProteinPilot or Mascot protein database search algorithms to identify the peptides and proteins (refer to Experiment 8).

REFERENCES

Karas M. and Hillenkamp F. 1988. Laser desorption ionization of proteins with molecular masses exceeding 10000 daltons. *Anal. Chem.* **60:** 2299–2301.

Karas M., Glückmann M., and Schäfer J. 2000. Ionization in matrix-assisted laser desorption/ionization: Singly charged molecular ions are the lucky survivors. *J. Mass Spectrom.* **35:** 1–12.

Mamyrin B.A., Karataev V.I., Shmikk D.V., and Zagulin V.A. 1973. The mass–reflectron, a new nonmagnetic time-of-flight mass spectrometer with high resolution. *Sov. Phys. JETP* **37:** 45–48.

Wolff M.M. and Stephens W.E. 1953. A pulsed mass spectrometer with time dispersion. *Rev. Sci. Instrum.* **24:** 616–617.

EXPERIMENT 4

Analysis of Protein Complexes
High-sensitivity Liquid Chromatography Coupled with Tandem Mass Spectrometry

Proteomics, or the large-scale analysis of proteins, has emerged as a result of the genome sequencing projects. The development of methods and instrumentation for automated tandem mass spectrometry (MS/MS) in combination with microcapillary liquid chromatography has dramatically increased the sensitivity and speed to identify proteins. Identification of proteins is now achieved with greater sensitivity than silver-stained gels. Furthermore, automated computer algorithms have supplemented manual interpretation of MS/MS spectra. Sophisticated programs are used to correlate tandem mass spectra of peptides with genomic sequences for rapid and conclusive protein identification. Because each mass spectrum is compared to all sequences in the database, the result is the unbiased identification of proteins. Using this approach, novel and unexpected proteins are being identified in a growing list of protein complexes, subcellular locations, and total cellular proteomes.

MS/MS is an extremely powerful technique for the sequence analysis of peptides in complex mixtures. In tandem mass spectrometric sequencing of a peptide, information about the sequence of the peptide is contained in the product ion or tandem mass spectrum. MS/MS experiments can be performed in a variety of mass spectrometers. Spatially separate mass analyzers (such as a triple quadrupole, quadrupole-time of flight [QTOF], or time of flight/time of flight [TOF/TOF]) and analyzers that separate ions in time (such as ion traps or Fourier transform ion cyclotron resonance [FTICR]) are predominantly used to sequence peptides by mass spectrometry. For peptide sequencing, the fragmentation spectrum is obtained by selecting a positively charged precursor ion from the survey or precursor mass spectrum, isolating the selected ion and ejecting all other ions, and finally fragmenting the precursor ion by collision induced dissociation (CID) to generate product ions. Low energy fragmentation of ionized tryptic peptides occurs primarily at the amide bonds along the peptide backbone, generating a series of fragmentation or product ions. Several models have been proposed that describe the fragmentation chemistry, including the mobile proton model and the pathways in competition model (Dongré et al. 1996; Wysocki et al. 2000; Paizs and Suhai 2005). The differences between product ions and the residue mass of the amino acids are used to determine the sequence of the peptide (see Experiment 8). The fragmentation patterns are also unique signatures for proteins in the sample that can be used by database search algorithms to identify the proteins.

Using electrospray ionization (ESI), liquid chromatography can be coupled directly with MS/MS (Fig. 1). This combination allows the mass spectrometer to analyze complex peptide mixtures. Peptides are separated by their chemical properties in the chromatography step and then separated by their *m/z* value in the mass spectrometer with subsequent sequence analysis by MS/MS. Computer methods allow data-dependent acquisition of tandem mass spectra in real time and the automated acquisition of MS/MS spectra. The instrument is programmed to choose which precursor ions to select or ignore for MS/MS analysis. Typically, the most abundant ions are selected for fragmentation, because they will usually produce MS/MS spectra with the strongest signals. During

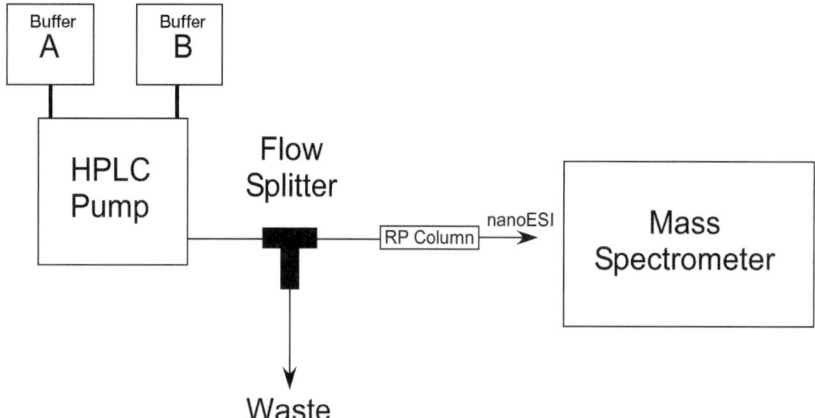

FIGURE 1. Diagram of the principal components of a nanoLC-MS/MS mass spectrometry system for proteomics.

a coupled liquid chromatography/tandem mass spectrometry analysis, it is possible to acquire a large number of MS/MS spectra: >4 MS/MS spectra every second using ion trap mass analyzers.

The genome sequencing projects have generated a wealth of information, including the theoretical sequences of all the proteins in a large number of organisms. A large number of computer programs have been developed to match uninterpreted tandem mass spectra to sequences in protein or nucleic acid databases. The programs compare the experimental spectra with theoretical spectra generated from protein databases and generate a list of peptides and proteins in the sample. Experiment 8 trains students to use specific search engines to process and analyze MS/MS data for identifying proteins and posttranslational modifications. Most importantly, the experiment instructs students in how to evaluate the accuracy of the peptide and protein identifications that the programs return.

In this experiment, microcapillary liquid chromatography is coupled to MS/MS (LC-MS/MS) to analyze complex protein mixtures. The goal is to comprehensively identify the proteins in the sample. In earlier experiments, protein complexes or mixtures were reduced and alkylated to denature the proteins and to derivatize the cysteine residues to prevent the formation of disulfide bonds. The protein complexes were digested with trypsin to cleave the proteins into peptides. This experiment analyzes those peptide mixtures using data-dependent microcapillary LC-MS/MS.

Microcapillary reversed-phase HPLC columns will be constructed for separating complex peptide mixtures. The microcapillary columns will be connected to an HPLC pump coupled to the tandem mass spectrometer using an ESI source. The HPLC pump and an ion trap mass spectrometer will be programmed for running a data-dependent LC-MS/MS experiment on the sample. The mass spectrometer will be first programmed to perform a precursor scan to measure the *m/z* values and intensities of ions eluting from the RP column. Second, the instrument will then be programmed to individually fragment (MS/MS) the most abundant ions using the data collected in the precursor scan. The ions to be fragmented are "dependent" upon the information in the precursor scan. The cyclic process of a precursor scan followed by a series of fragmentation or MS/MS scans is programmed to constantly repeat during the entire LC gradient.

After adjusting the HPLC's mobile phase flow rate through the column, the liquid chromatography and mass spectrometer will be evaluated by loading a control peptide, angiotensin I, and running an LC-MS/MS experiment. After verifying the performance and sensitivity of the system, the experimental protein samples from other experiments will be analyzed by LC-MS/MS. In all of the experiments, the samples will be manually loaded onto the microcapillary column and analyzed. When the LC-MS/MS is finished, the data files will be transferred to a computer system running database search programs and the data processed and analyzed to generate a list of peptides and proteins (Experiment 8).

PROTOCOL 1

Making Microcapillary HPLC Columns

In ESI of peptides, an acidic aqueous solution that contains the peptides flows through a small-diameter needle. A high, positive voltage is applied to the needle to produce a Taylor cone as the solution exits the needle. Small droplets of solution are generated by the Taylor cone, which contains the peptide analyte. Protons from the acidic solution give the droplets a positive charge, causing them to move from the needle to the negatively charged instrument. During the movement, evaporation reduces the size of the droplets and the droplets split into smaller and smaller, charged droplets. The evaporation and splitting process eventually causes the peptides to desorb into the gas phase as protonated peptides. The ionized peptides can be directed and manipulated by the mass spectrometer's electric or magnetic fields.

Reversed-phase (RP) chromatography fractionates peptides and proteins based on the interaction between hydrophobic patches on the surface of biomolecules and nonpolar alkyl chains bonded covalently to the surface of the stationary phase. RP chromatography is used with ESI because RP's acidic aqueous and polar mobile phases are compatible with ESI. In addition, in-line RP-HPLC is useful for desalting peptides before ESI without the need for off-line desalting steps. RP-HPLC tends to focus peptides from dilute samples into narrow chromatographic bands, which enhances sensitivity.

The use of an acidic solution tends to protonate all the available basic residues in a peptide. These include the amino-terminal amine and the basic side groups of lysine (K), arginine (R), and histidine (H). As a result, multiply-charged protonated peptides are observed when a peptide contains a K, R, or H residue. Since trypsin cleaves peptides at the carboxy-terminal side of R and K, those tryptic peptides tend to be doubly charged (amino-terminal amine and the K or R residue). Tryptic peptides containing internal basic residues (i.e., internal H, R-P, K-P, or cleavage sites that trypsin missed) are typically more highly charged peptides (e.g., +3, +4, etc.).

The sensitivity of ESI-LC-MS/MS is inversely proportional to the flow rate. The low flow rates (<0.5 µL/min) of microcapillary HPLC are orders of magnitude more sensitive than standard RP-HPLC columns with flow rates of 50 µL/min or more. For the successful analysis of low femtomole (nanogram) amounts of peptides, microcapillary HPLC is required.

This protocol describes the construction of a pulled microcapillary column containing an approximately 3-µm orifice at the end of a fused-silica capillary (FSC) (Fig. 2). A laser-based micropipette puller is used to pull the capillary and create the restriction. The restriction prevents packing material from passing through the column but allows liquid to flow through. The pulled tip also functions as the emitter tip for ESI. For ESI-LC-MS/MS, the integrated column and emitter needle are connected to an electrospray interface. In Protocol 2, the FSC is packed with RP packing material. These protocols will be used later (Experiment 6) to construct microcapillary columns for multidimensional peptide separations for 2D LC-MS/MS experiments or MudPIT.

If you prefer not to construct individual columns or do not have access to a laser puller, packed 1D and 2D fused-silica microcapillary columns can be purchased from commercial vendors.

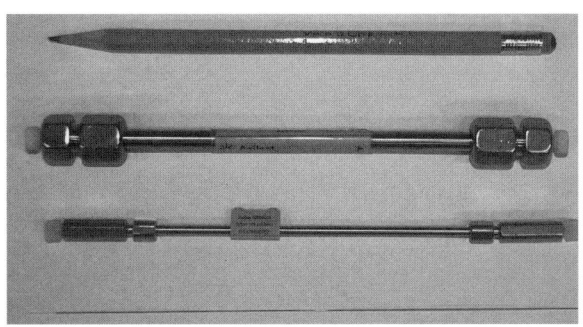

FIGURE 2. Different size RP-HPLC columns used for LC-MS/MS experiments. For RP-HPLC, a 4.6-mm, 1-mm, and 0.1-mm inner diameter column can be used. For high-sensitivity proteomics LC-MS/MS experiments, the 0.1-mm or 100-µm fused silica capillary (FSC) column, shown at the bottom of the figure, is typically used.

MATERIALS

CAUTION: See Appendix 11 for appropriate handling of materials marked with <!>.

Reagents

Methanol <!>

Equipment

Alcohol lamp
Fused silica capillary scribes (Chromatography Research, 205312)
100 µm ID x 365 µm OD Fused silica capillary (FSC) tubing (PolyMicro Technologies)
Laser-based micropipette puller (e.g., P-2000 Sutter Instruments)

PROCEDURE

1. Cut ~18 inches of 100 µm ID x 365 µm OD FSC with a cleaving tool (Fig. 3A,B).

2. Burn a 1–2-inch window in the middle of the capillary with an alcohol lamp. Slowly rotate the FSC over the lit flame (Fig. 4).

   Only heat the FSC until the polyimide coating has been charred. Excessive heating will damage the capillary.

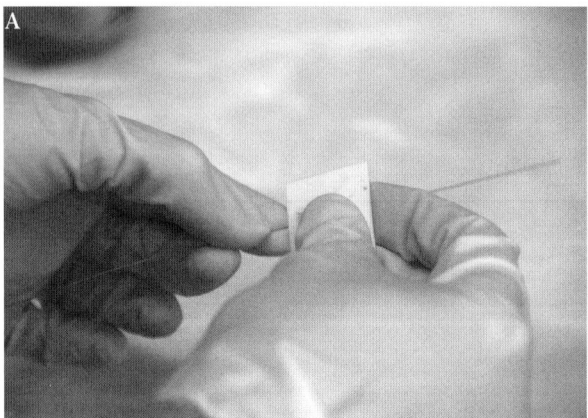

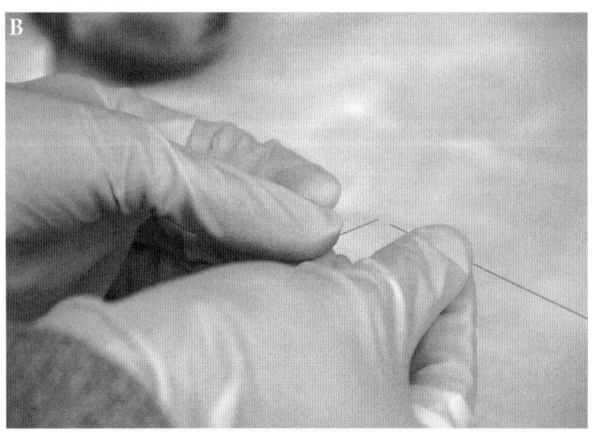

FIGURE 3. Cleaving fused silica capillary (FSC). (A) A silica scribe is used to lightly score the FSC. (B) After the plastic coating and quartz glass are scored with the silica scribe, the FSC will break into two pieces with a light force.

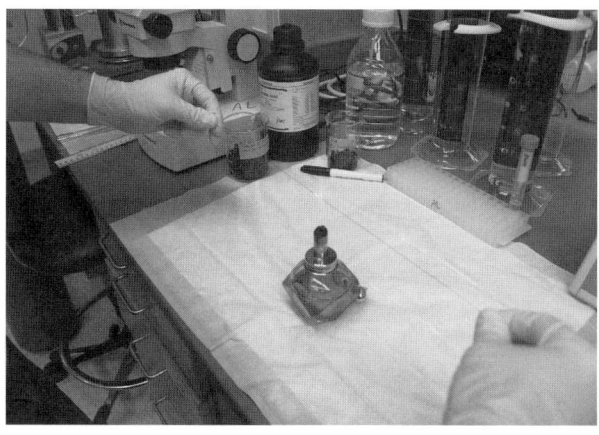

FIGURE 4. Using an alcohol lamp to burn off the polyimide plastic coating from the FSC. The FSC is briefly held in the flame until the coating turns black.

3. Using a Kimwipe wet with methanol, gently wipe the charred polyamide coating off to expose the clear quartz (Fig. 5A,B).

   Be careful as the quartz is very fragile and breaks easily. Be sure to remove all charred pieces off the quartz.

4. Place the capillary in the Sutter P-2000 needle puller and align the capillary using the grooves (Fig. 6A,B). Tighten the FSC down with the clamps before lowering the lid. Use the program on the next page to pull two columns.

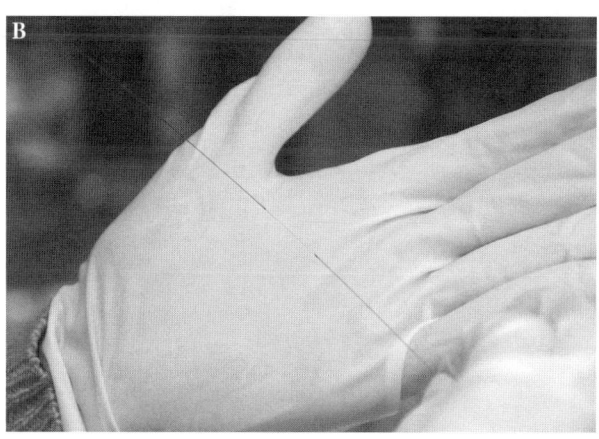

FIGURE 5. (A) Using a methanol-soaked Kimwipe, gently wipe clean the charred black plastic coating from the FSC to expose the quartz glass. (B) Exposed quartz glass of an FSC ready for pulling into microcapillary HPLC columns using the laser puller.

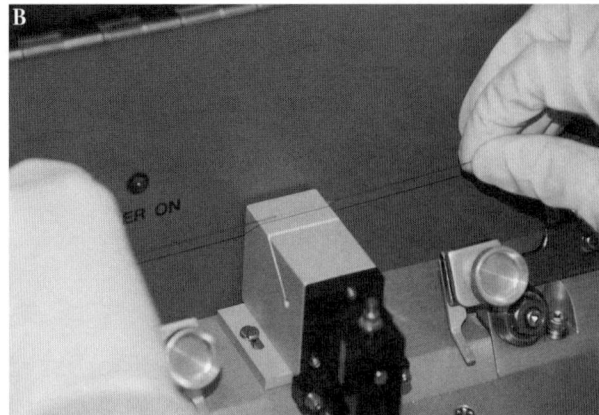
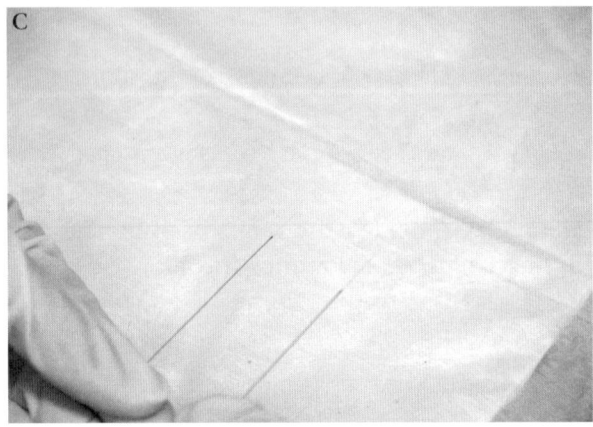
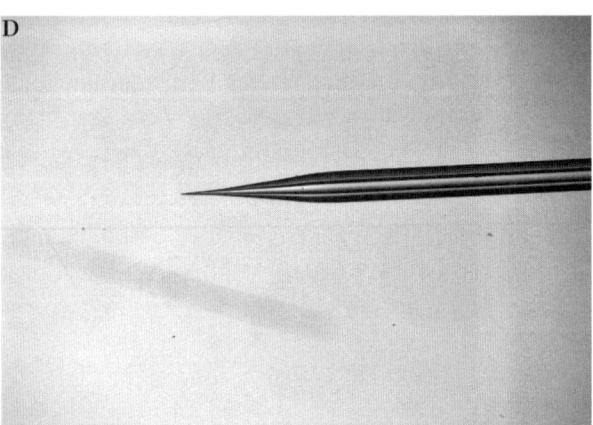

FIGURE 6. (A) Sutter laser micropipette puller for making microcapillary HPLC columns for nanoLC-MS/MS experiments. (B) Placing the exposed FSC in the laser puller. The exposed, cleaned quartz glass is centered in the middle of the laser. (C) Two pulled, empty FSC microcapillary columns that are ready for packing. (D) Close-up of the tip created in the FSC using the laser puller. The symmetrical tip is necessary for generating a stable nanoelectrospray.

Heat	Velocity	Delay
320	40	200
310	30	200
300	25	200
290	20	200

If the red laser light flashes and the puller's jaws separate, you should have a successful pull. If the light does not flash, press the stop button and try to align the FSC again. A successful pull will create two fused silica capillary columns, each with a restriction at one end (Fig. 6C,D). The parameters used for the puller program may need to be adjusted to account for instrument variation.

PROTOCOL 2

Packing Microcapillary FSC Columns

MATERIALS

CAUTION: See Appendix 11 for appropriate handling of materials marked with <!>.

Reagents

5% Acetonitrile <!>, 0.1% formic acid <!>
Methanol <!>
Reversed-phase resin (Phenomenex Synergi 4u Hydro-RP 80A)

Equipment

Glass vials with screw caps, 1.8-mL (Chromatography Research 123315/309925)
Micro stir bars (VWR 58948-069)
Microcapillary column (from Protocol 1)
Pneumatic packing vessel (homemade or commercially available from New Objectives or Next Advance, Inc.)
Ruler
Sonicator <!>
Stereomicroscope
Stir plate
Vortexer

PROCEDURE

1. Place a microstir bar into a 1.8-mL glass vial. Add 0.6 mL of methanol and 8 mg of RP packing material to the glass vial.

   The thickness of the slurry will determine how quickly the column will pack.

2. Vortex to resuspend the packing material and sonicate for 5 minutes to prevent aggregation of the particles.

3. Transfer the slurry to a pneumatic loading vessel and place the loading vessel on a stir plate (Fig. 7A,B,C). Turn on the stir plate to keep the packing material suspended. Secure the lid to the loading vessel by tightening the bolts that attach the lid to the base.

   WARNING: Always wear safety glasses when packing FSC columns. The columns are being packed under very high pressure. Improperly seated columns can be ejected from the loading vessel at high velocity.

4. Measuring from the frit end, place a mark on the empty microcapillary column showing the desired packing height.

   For a 100 µm ID x 365 µm OD pulled microcapillary column, 9 cm of RP material is packed in the column.

5. Feed the empty microcapillary column down through the Vespel ferrule in the Swagelok fitting on the lid of the pneumatic loading vessel until the end reaches the bottom of the vial. Pull the column up so that the capillary rests just off the bottom of the glass vial and stir bar, and tighten the ferrule to secure the column (Fig. 8).

6. Apply pressure to the loading vessel by first setting the regulator on the high-pressure helium gas cylinder to 500–1000 psi, then opening the three-way valve.

7. If the column is long enough, place the fritted end of the column under a stereomicroscope to observe the packing. Be extremely careful as the column tip is very fragile.

   A steady stream of packing material should be seen flowing into the capillary. For pulled columns, the tip may need to be opened slightly when packing the column. To do this, gently score the opening of the column (Fig. 9A,B). In a smooth upward motion, glide the capillary

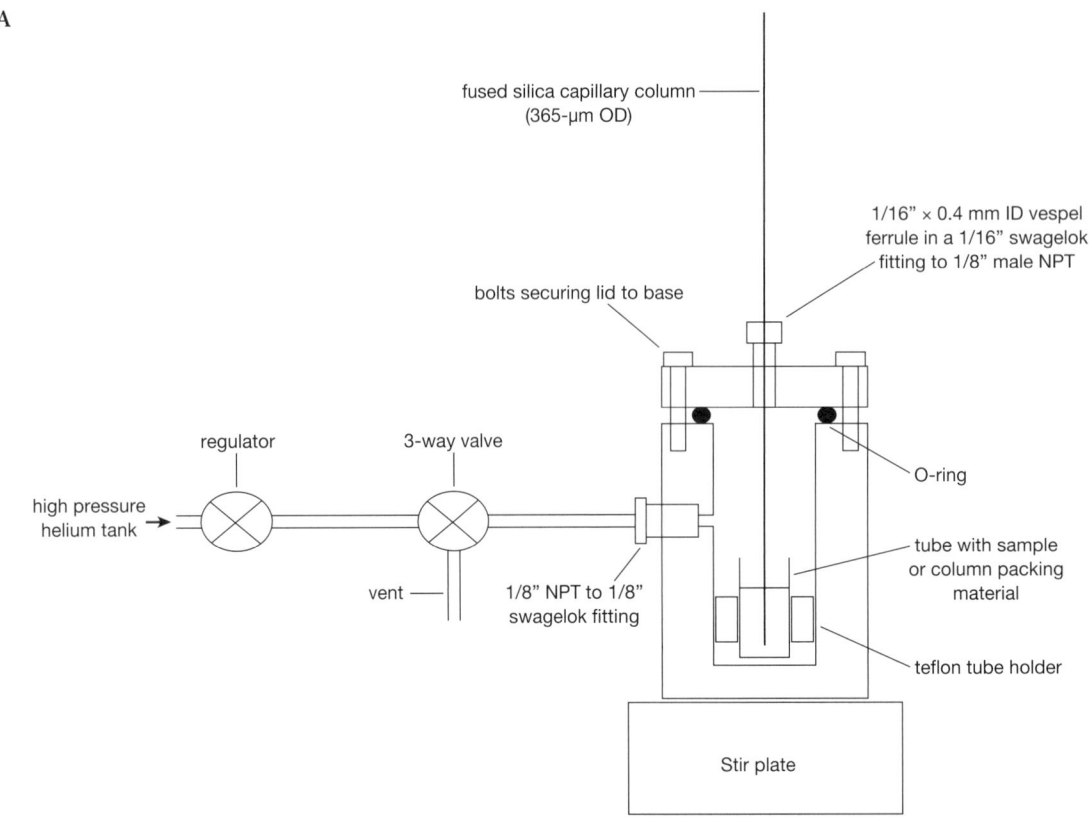
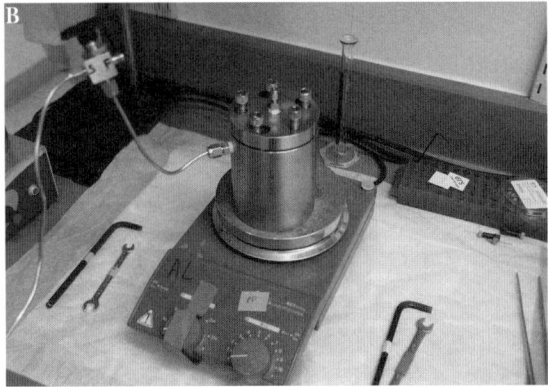

FIGURE 7. (A) Diagram of the pneumatic loading device, "loading bomb," for packing microcapillary FSC HPLC columns. (B) Actual loading bomb on a stir plate. (C) Using tweezers to insert a glass vial containing the methanol-reverse phase resin slurry for packing the FSC column. The vial contains a microstir bar for keeping the slurry in suspension while the column is packing.

cleaving tool along the side of the FSC. Never score directly on top of the column since it will destroy the tip for electrospray (Fig. 10).

8. When the column has been packed to the mark, slowly turn off the pressure to the vessel at the three-way valve.

    Slowly releasing the pressure prevents the packing material from unpacking.

9. Replace the vial containing the slurry with a 1.5-mL microcentrifuge tube filled with 5% acetonitrile, 0.1% formic acid. Wash the column for 10 minutes using the loading vessel.

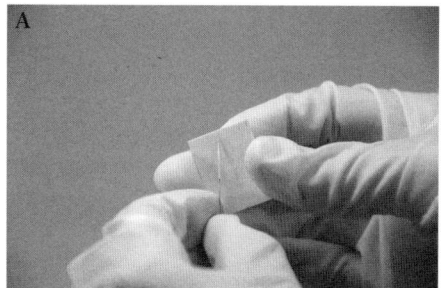

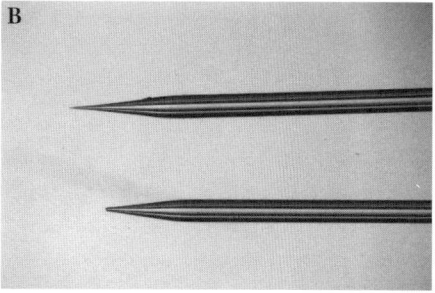

FIGURE 8. Empty FSC column inserted into the loading bomb for packing.

FIGURE 9. (A) Using a silica scribe to "open" the pulled FSC column. To get the columns flowing, it is typically necessary to lightly score the side of the tip. (B) Examples of unscored and properly scored FSC tips.

10. Store the column in 5% acetonitrile, 0.1% formic acid until ready to use. Columns can be stored indefinitely at room temperature in this solution.

    The column must be completely submerged in 5% acetonitrile, 0.1% formic acid to prevent the packing resin from drying out.

11. Before using for any biological samples, run a blank HPLC gradient across the column to condition it.

    Conditioning the column is important to firmly pack the resin. Multiple blank HPLC gradients may be required to obtain a reproducible baseline. If high sensitivity applications are planned, load 0.1 pmol of angiotensin peptide onto the new column and run an HPLC gradient (see Protocol 3). Angiotensin binds to nonspecific binding sites in the column and minimizes nonspecific, irreversible binding of sample peptides.

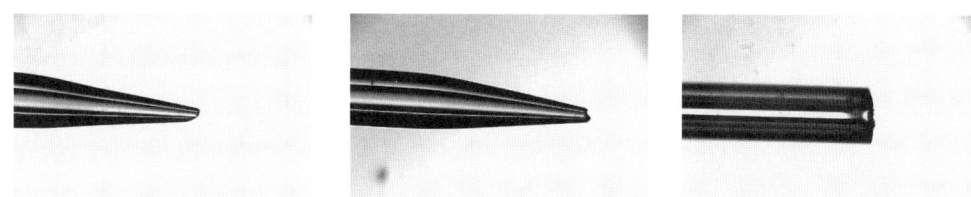

FIGURE 10. Examples of FSC column tips that have been excessively scored to get the packing slurry flowing.

PROTOCOL 3

Microcapillary RP-HPLC Coupled to ESI-Mass Spectrometry

This protocol describes the RP-LC-ESI-MS assembly and LC-MS/MS process.

MATERIALS

CAUTION: See Appendix 11 for appropriate handling of materials marked with <!>.

Reagents

Acetonitrile <!> (for HPLC gradients; see Step 13)
Angiotensin solution (0.02 pmol/µL) (angiotensin I, Sigma-Aldrich A9650) or Trypsin-digested protein sample (see Step 4)
Solvent A (5% acetonitrile, 0.1% formic acid <!>)

Equipment

Disposable calibrated glass pipettes, 5 µL (Drummond Scientific 2-000-001)
50 µm ID x 365 µm OD FSC tubing (PolyMicro Technologies)
75 µm ID x 365 µm OD FSC tubing (PolyMicro Technologies)
Fritless microcapillary RP column (from Protocol 2)
Fused silica capillary scribe (Chromatography Research 205312)
HPLC pump (model 1200; Agilent)
Linear ion trap mass spectrometer (model LTQ; Thermo Scientific)
Nanospray ESI source (James Hill Instruments)
PEEK MicroTee (Upchurch P775)
PEEK 380 µm ID MicroTight sleeve
Pneumatic loading device

PROCEDURE

1. Prepare the RP-apparatus as follows (Fig. 11A,B).

   a. Connect the transfer line from the HPLC pump to the arm of a PEEK-restrictor MicroTee using a PEEK 380 µm ID MicroTight sleeve.

   b. Connect the PEEK-restrictor MicroTee from the center arm to a PEEK-ESI MicroTee using a 75 µm ID x 365 µm OD FSC and PEEK sleeves.

   c. Connect a 30-cm piece of 50 µm ID x 365 µm OD FSC restrictor line through the third arm of the PEEK-restrictor Tee using a PEEK sleeve.

   d. Connect a 0.025-inch OD gold wire through the center arm of the PEEK-ESI Tee and attach it to the ESI voltage source.

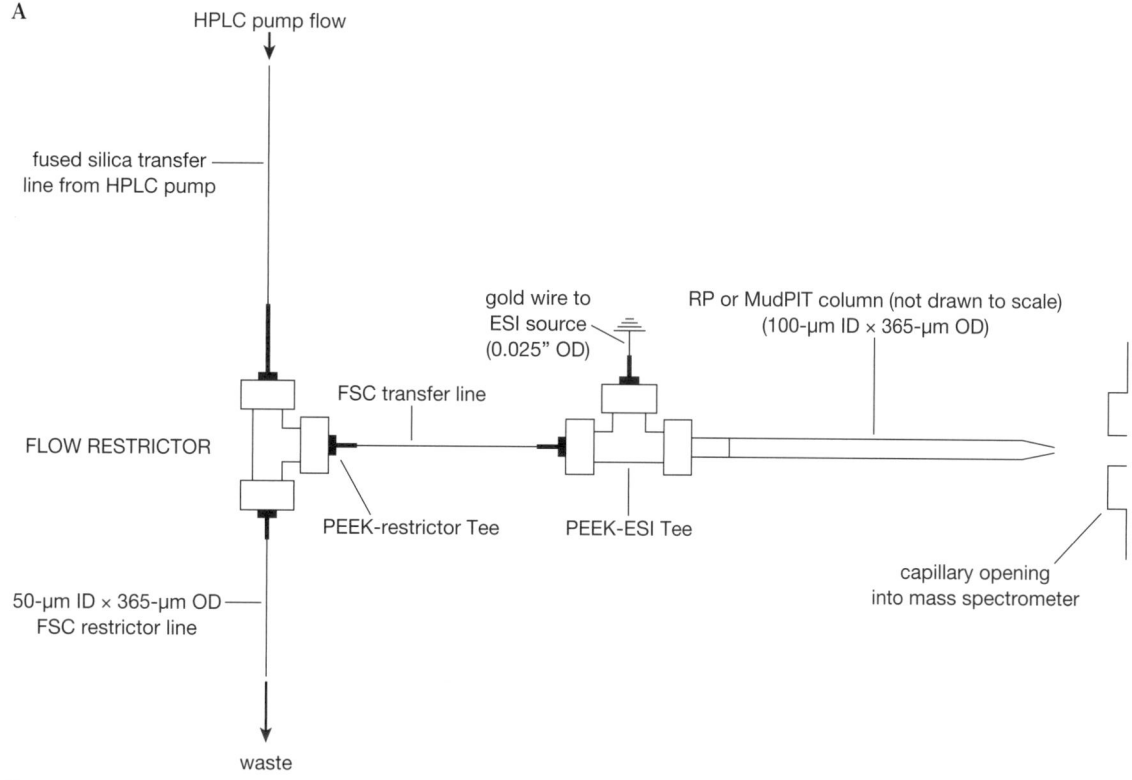
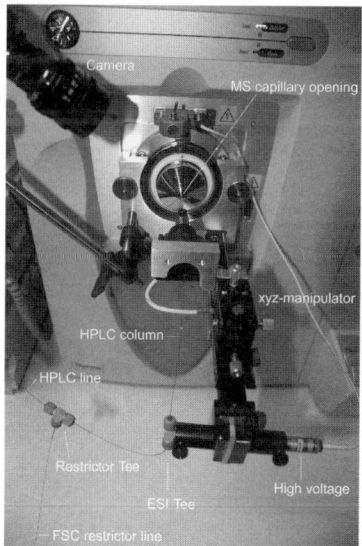

FIGURE 11. (A) Diagram of the FSC connections used to construct the nanoESI source. (B) Actual nanoESI source mounted to an ion trap mass spectrometer. The video camera is used to assist in the alignment of the FSC HPLC column tip with the mass spectrometer's heated capillary tube opening.

  e. Connect the pulled microcapillary RP column to an arm of the Tee using a PEEK-ESI Tee.

   The pulled microcapillary HPLC RP column assembly is mounted to an x-y-z manipulator at the entrance of the mass spectrometer (Figs. 11 and 14). The manipulator allows fine adjustment of the column tip with respect to the mass spectrometer's capillary opening (see Step 12). The pulled column tip is extremely fragile. Avoid letting the tip strike a solid surface. A 2.2 kV voltage is applied to the gold wire during ESI.

2. Set the HPLC pump to 100% solvent A with a flow rate of 200 µL/min.

FIGURE 12. Measuring the flow rate through the microcapillary FSC HPLC column using a calibrated glass capillary pipette.

3. Measure the flow rate through the column for 1 minute using a 5-µL calibrated glass capillary pipette (Fig. 12). Trim the 50 µm ID x 365 µm OD restrictor line using the capillary cleaving tool until a flow rate of 0.2 to 0.5 µL/min through the RP column is obtained (see Fig. 11A).

   Measurement of the flow rate and adjustment of the split line may have to be repeated several times until the target flow rate is achieved. Several vendors now offer specialized HPLC pumps with low mobile phase flow rates (<1 µL/min) that generate reproducible gradients. The nanoflow pumps eliminate the need for the external PEEK-restrictor Tee and flow splitting.

4. Place a tube containing the trypsin-digested protein sample or the angiotensin solution into a pneumatic loading vessel and tightly attach the lid of the vessel.

   To condition the microcapillary column and measure the performance of the HPLC system and mass spectrometer before analyzing unknown samples, a control sample is typically first run before analyzing one's precious biological sample. If time does not permit the control experiment, the RP column can be conditioned off-line and the unknown sample loaded onto the column. Angiotensin I (DRVYIHPFHL) has a monoisoptic and average mass of 1295.68 and 1296.49 gm/mol, respectively. During ESI, it predominantly forms a +3 ion at approximately 433 m/z. Monitoring the retention time (RT), signal intensity, and resolution of the angiotensin peptide (and background noise) is used to check the performance of the chromatography and mass spectrometry system prior to running unknown samples. Alternative control peptides or trypsin-digested control proteins (e.g., BSA) can be used to evaluate the system.

5. Disconnect the column from the PEEK-ESI Tee assembly and insert the open end of the column into the top of the loading vessel until the capillary reaches the bottom of the sample tube. Pull the column up slightly off the bottom of the sample tube.

   Leaving a small gap between the capillary and the bottom of the tube reduces the probability of any solid particulates at the bottom of the tube clogging the capillary column.

6. Secure the column to the bomb by tightening the compression nut on the Swagelok fitting at the top of the loading vessel.

   Tightening the Swagelok fitting's nut compresses the Vespel ferrule around the FSC.

7. Apply pressure to the loading vessel by first setting the regulator on the helium gas cylinder to 500–1000 psi, and then opening the three-way valve.

8. Measure the volume loaded using a 5-µL calibrated glass pipette to collect the displaced volume from the end of the column (see Fig. 13).

   For the angiotensin I control sample, load 5 µL (0.1 pmol) of the angiotensin solution. For unknown samples, the amount of sample to load can be problematic. If low amounts of unknown sample are available (silver-stained bands), loading the entire sample is recommended. For more abundant samples (Coomassie-stained bands), we typically load a fraction of the sample.

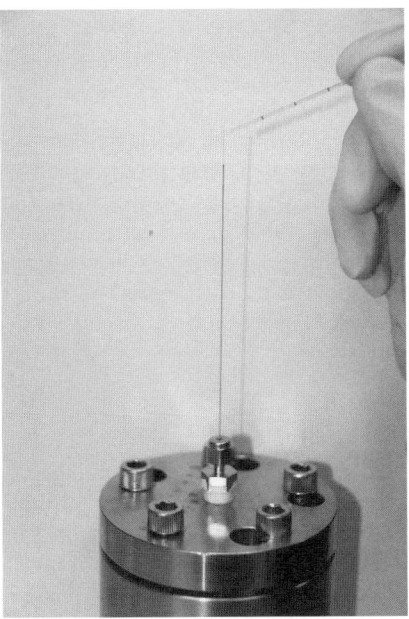

FIGURE 13. Using the bomb to load a sample onto the microcapillary FSC HPLC column. A microcentrifuge tube with the sample is placed in the loading bomb and the column is inserted. When the bomb is pressurized, a calibrated glass capillary pipette is used to measure the amount of sample loaded onto the column.

9. Release the pressure in the loading vessel using the three-way valve once the sample is loaded.

10. Reinstall the column in the union.

    The RP column will be plumbed for performing microspray mass spectrometry on the angiotensin or an unknown sample as it elutes from the column.

11. Start the HPLC flowing at 200 μL/min with 100% Solvent A and re-check the RP column flow rate using a 5-μL calibrated glass pipette.

12. Carefully position the pulled microcapillary HPLC column tip at the entrance of the mass spectrometer (Fig. 14). Use a camera monitor to assist in positioning the column at the optimal position.

    The pulled microcapillary column tip is centered 1–5 mm from the orifice of the capillary opening into the mass spectrometer using the x-y-z manipulator with the aid of the closed circuit monitor (Fig. 14). The pulled column tip is extremely fragile. Do not let the tip strike a solid surface.

13. Program the HPLC using the following as an example: 60-minute gradient from 0% to 40% acetonitrile, and a 10-minute gradient from 40% to 60% acetonitrile (see Appendix 7).

    There are many gradient variations and RP buffers that can be used.

FIGURE 14. Position of the microcapillary HPLC column tip relative to the opening of the mass spectrometer's ion transfer tube.

14. Using a Kimwipe, carefully wick away any large drops from the column tip. Start the mass spectrometer and the HPLC gradient and collect MS/MS data on peptides as they elute from the RP column.

    A voltage of ~2.2 kV is applied to the gold wire during ESI. For the ThermoFisher LTQ mass spectrometer, data-dependent acquisition of tandem mass spectra is programmed through the instrument's Xcalibur software. Typical data acquisition settings consist of a continual cycle beginning with one MS scan with an *m/z* scan range of 300–2000, which records all the *m/z* values of ions present at that moment in the gradient, followed by five data-dependent MS/MS scans. The MS/MS scans fragment the five most abundant ions recorded in the first MS scan. Dynamic exclusion is activated to improve protein identification capacity by avoiding the repeated fragmentation of abundant ions. Each instrument manufacturer has different options for configuring tandem mass spectra collections.

15. When the ESI-LC-MS/MS run is complete, turn off the ESI voltage.

    Examine the acquired data for angiotensin peptide to confirm the instrument is properly functioning before analyzing unknown samples. The angiotensin retention time is typically 33–36 minutes. For the LTQ instrument, a signal intensity >10E7 for a 433 ion should be observed. These values are typical for a properly functioning HPLC and linear ion trap LTQ mass spectrometry system. As described in Experiment 8, the acquired data file can be searched against a protein database containing the angiotensin I sequence to confirm the MS/MS. If these values or protein identifications are not observed, a number of steps can be taken, including cleaning the ESI source, checking the HPLC flow rates, and re-calibrating and retuning the mass spectrometer. Other HPLC and mass spectrometry systems will have different RT and signal intensities. Practical experience running an LC-MS/MS system is the best way to learn how to operate the instrumentation and to troubleshoot problems. There is no substitute for practical, hands-on experience.

16. Equilibrate the RP column for 10 minutes at 100% Solvent A before loading unknown samples from other experiments in this manual.

17. Load and analyze the unknown sample onto the conditioned and tested microcapillary RP column and mass spectrometry system using Steps 4–15.

    For LC-MS/MS analysis of a purified protein complex, Steps 4–15 are repeated, except that the unknown, trypsin-digested biological samples are now loaded onto the RP column and analyzed by LC-MS/MS. Because of possible carryover on the RP column, it is common to run a blank between unknown samples to check that the tryptic-digested proteins have completely eluted.

18. Process and analyze the acquired mass spectrometry data file as described in Experiment 8 to identify the peptides and proteins.

REFERENCES

Dongré A.R., Jones J.L., Somogyi A., and Wysocki V.H. 1996. Influence of peptide composition, gas-phase basicity, and chemical modification on fragmentation efficiency: Evidence for the mobile proton model. *J. Am. Chem. Soc.* **118:** 8365–8374.

Link A.J., Jennings J.L., and Washburn M.P. 2003. Analysis of protein composition using multidimensional chromatography and mass spectrometry. In *Current protocols in protein science* (ed. J.E. Coligan et al.), chapter 23, pp. 1–25. John Wiley and Sons, New York.

Paizs B. and Suhai S. 2005. Fragmentation pathways of protonated peptides. *Mass Spectrom. Rev.* **24:** 508–548.

Wysocki V.H., Tsaprailis G., Smith L.L., and Breci L.A. 2000. Mobile and localized protons: A framework for understanding peptide dissociation. *J. Mass Spectrom.* **35:** 1399–1406.

EXPERIMENT 5

Phosphopeptide Analysis Using IMAC and Mass Spectrometry*

While protein identification by mass spectrometry has become routine, protein phosphorylation analysis remains a challenging problem. Low phosphorylation stoichiometry, heterogeneous phosphorylation sites, and low protein abundance contribute to the difficulty of phosphoprotein analysis. In addition, phosphopeptides are generally difficult to analyze by mass spectrometry (Mann et al. 2002). Reduced ionization efficiency and suppression by nonphosphorylated peptides exacerbate the problems when analyzing phosphopeptides by mass spectrometry. To improve the efficiency of phosphopeptide analysis by LC-MS/MS, various strategies have been developed to enrich for phosphopeptides, including strong cation exchange chromatography (SCX), immunoaffinity capture using antiphosphotyrosine antibodies, and immobilized metal affinity chromatography (IMAC) (Andersson and Porath 1986; Posewitz and Tempst 1999; Beausoleil et al. 2004; Pinkse et al. 2004; Larsen et al. 2005; Rush et al. 2005).

IMAC exploits the affinity of the negatively charged phosphorylated amino acids (phospho-Ser, -Thr, and -Tyr) for positively charged metal ions. The metal ions (e.g., Fe^{3+}, Ga^{3+}, TiO_2) are chelated to stationary-phase chromatographic media. Liquid chromatography is then used to enrich for phosphopeptides. IMAC was originally plagued by the nonspecific retention of peptides rich in acidic amino acids. Recent improvements in IMAC methodology have increased its specificity and sensitivity (Ficarro et al. 2002, 2005; Moser and White 2006). To minimize the binding of nonphosphorylated peptides to an IMAC column, tryptic peptides from digested cell extract are converted to the peptide methyl esters. The esterification (+14 Da) of the carboxyl groups at the carboxyl terminus and the glutamic and aspartic amino acids neutralizes their negative charge. Peptide esterification and IMAC have been successfully used to enrich and identify large numbers of phosphopeptides in yeast and human carcinoma cell lines (Ficarro et al. 2002; Kim et al. 2005). IMAC has also been used to enrich for phosphotyrosine peptides after protein immunoprecipitation from human cell lines (Brill et al. 2004). The direct coupling of IMAC with high-sensitivity reversed-phase liquid chromatography and tandem mass spectrometry has enabled researchers to identify phosphopeptides from limited amounts of material. In this experiment, peptide esterification and IMAC are used to enrich for phosphopeptides from whole-cell lysates (Fig. 1). The captured phosphopeptides are identified using LC-MS/MS (see Experiments 4 and 8).

---

*Protocols contributed by Forest M. White (*Department of Biological Engineering, Massachusetts Institute of Technology, Cambridge, Massachusetts 02139*), Paul H. Huang (*Department of Biological Engineering, Massachusetts Institute of Technology, Cambridge, Massachusetts 02139*), and Adam R. Farley (*Department of Microbiology and Immunology, Vanderbilt University School of Medicine, Nashville, Tennessee 37232*).

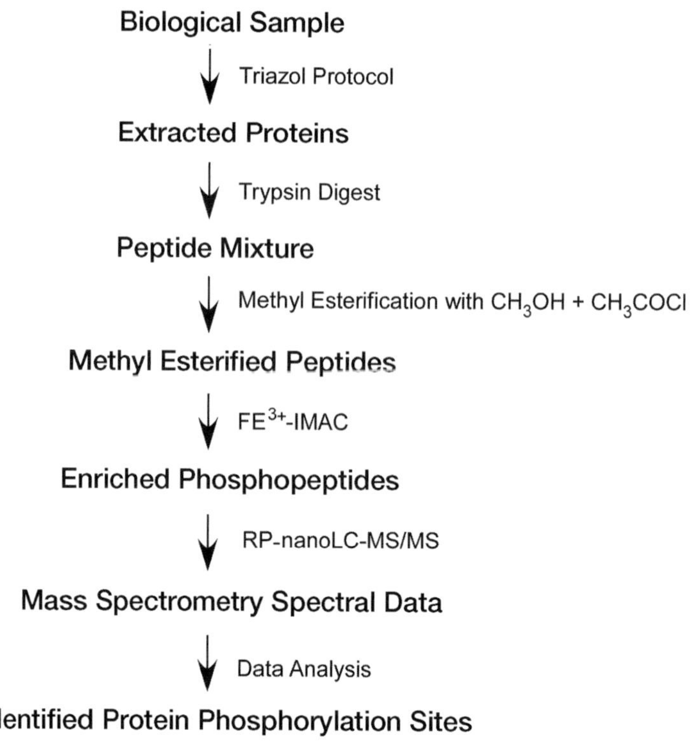

Biological Sample
↓ Triazol Protocol
Extracted Proteins
↓ Trypsin Digest
Peptide Mixture
↓ Methyl Esterification with $CH_3OH + CH_3COCl$
Methyl Esterified Peptides
↓ FE^{3+}-IMAC
Enriched Phosphopeptides
↓ RP-nanoLC-MS/MS
Mass Spectrometry Spectral Data
↓ Data Analysis
Identified Protein Phosphorylation Sites

FIGURE 1. Flowchart of the phosphoproteomics experiment.

PROTOCOL 1

Preparing Sample for IMAC

MATERIALS

CAUTION: See Appendix 11 for appropriate handling of materials marked with <!>.

Reagents

100 mM Ammonium acetate (pH 8.9)
Chloroform <!>
100% Ethanol <!>
100% Isopropanol <!>
Methanol, anhydrous (Sigma-Aldrich 322415) <!>
Rinse solution (0.3 M guanidine-HCl <!> in 95% ethanol)
1% SDS <!>
Thionyl chloride (Sigma-Aldrich 230464) <!>
Tissue or cell samples
Trizol reagent (Invitrogen 15596-018) <!>
Trypsin, modified, sequencing-grade (Promega) <!>

Equipment

Chopping/grinding device (see Step 13)
Homogenizer, glass-Teflon or power type

Sonicating water bath
Spatula, metal
Vacuum evaporator (e.g., SpeedVac; Savant)
Vial, glass
Water bath or heat block set to 37°C

PROCEDURE

Proteins to be analyzed can be obtained from either cell lines or tissue samples. The CSHL course uses a Trizol protocol from Invitrogen to prepare protein samples for analysis. There are a number of other protein isolation protocols that can used to isolate proteins from cells and tissues.

Preparation of Tryptic Digests

1. For tissue samples, homogenize 50–100 mg of tissue per mL of Trizol reagent using a glass-Teflon or power homogenizer. For cells grown in a monolayer, lyse cells directly in the culture dish by adding 1 mL of Trizol per 10 cm^2 of culture plate area (e.g., 1 mL of Trizol for a 3.5-cm diameter dish). Pass the cell lysate several times through a 1-mL pipette. For cells grown in suspension, pellet the cells by centrifugation. Lyse the cells by repetitively pipetting 1 mL of Trizol reagent per 0.5–1 x 10^7 cells.

    Use clear polypropylene tubes when working with Trizol.

2. Incubate the homogenized samples for 5 minutes at room temperature to completely dissociate protein complexes.

3. Add 0.2 mL of chloroform per mL of Trizol. Cap the tubes and shake vigorously for 15 seconds and incubate the samples at room temperature for 2 minutes.

4. Centrifuge the samples at 10,000g for 15 minutes at 4°C.

5. Completely remove the aqueous phase overlaying the interphase and organic phase.

    The aqueous phase contains the cellular RNA. The RNA can be precipitated from the aqueous phase by mixing with isopropyl alcohol. The interphase and organic phase contain the cellular DNA and proteins.

6. To the interphase and organic phase, add 0.3 mL of 100% ethanol per mL of Trizol used for the initial homogenization. Mix the samples by inversion and incubate the samples for 3 minutes at room temperature.

    The ethanol precipitates the cellular DNA from the organic phase.

7. Centrifuge the samples at 2,000g for 5 minutes at 4°C.

    The precipitated pellet contains the cellular DNA.

8. Transfer the Trizol-chloroform-ethanol supernatant to a fresh tube.

9. Precipitate the proteins from the Trizol-chloroform-ethanol supernatant with 1.5 mL of isopropanol per mL of Trizol used for the initial homogenization.

10. Incubate the samples for 10 minutes at room temperature. Centrifuge the protein precipitate at 12,000g for 10 minutes at 4°C.

11. Remove the supernatant and wash the protein pellet in rinse solution. Add 2 mL of rinse solution per mL of Trizol used for the initial homogenization. Mechanically grind the pellet with a metal spatula to break up the pellet. Sonicate the resuspended protein pellet for 5 minutes.

12. Centrifuge the protein precipitate at 12,000g for 10 minutes at 4°C. Decant the supernatant from the resulting pellet.

13. Mechanically chop/grind the pellet again and repeat Steps 11 and 12.

    The pellet is harder to dissociate after the first rinse.

14. Repeat Steps 11–13 until all of the pink color disappears from the protein pellet.

    The pellet should be a fine, white precipitate.

15. Finally, rinse the pellet with 100% ethanol, centrifuge, decant the ethanol, and resuspend the pellet in 100 µL of 1% SDS.

    Add additional 1% SDS solution if the pellet fails to resolubilize.

16. Dilute the resolubilized protein solution 6x with 100 mM ammonium acetate (pH 8.9). Add 20–40 micrograms of modified, sequencing-grade trypsin. Digest overnight at 37°C.

17. Dry the digested proteins completely in a vacuum evaporator.

CRITICAL: The peptides must not contain any H_2O for the methyl ester modification reaction. Samples can be frozen at this stage prior to conversion to methyl esters.

Esterification of Peptides

18. Working in a chemical fume hood, mix 1 mL of anhydrous methanol with 50 µL of thionyl chloride in a glass vial.

    Thionyl chloride is very reactive and should be slowly added drop-wise. Thionyl chloride will generate chlorine gas and a considerable amount of heat. Wear gloves and protective eye glasses. Anhydrous methanol and thionyl chloride should be stored in a vacuum dessicator to prevent H_2O contamination (It may be necessary to add molecular sieves to the anhydrous methanol prior to the reaction).

19. Add the entire methanol/thionyl chloride solution to the trypsin-digested proteins from Step 17. Sonicate for 15 minutes at room temperature in a sonicating water bath. Incubate for 2 hours at room temperature.

20. Lyophilize the peptides to dryness in a vacuum evaporator.

    The dried peptides can be stored at –80°C. The lyophilized peptides will be resuspended prior to loading onto the IMAC column in Protocol 4 of this experiment.

PROTOCOL 2

Construction of an IMAC Column and a Reversed-phase Capture Column (Precolumn)

MATERIALS

CAUTION: See Appendix 11 for appropriate handling of materials marked with <!>.

Reagents

Acetonitrile (HPLC grade) <!>
Formamide <!>

Kasil #1 potassium silicate solution (PQ Corporation)
   A 1-quart sample of Kasil #1 is free. It can be requested online from the PQ Corporation at http://www.pqcorp.com/corporate/samplerequest.asp
POROS MC 20 IMAC chromatography medium (Applied Biosystems 1-5429-06)
YMC-Gel ODS-A 12 nm S-10 μm reversed-phase medium (Kanematsu Corp AA12S11)
   Prepare as a slurry in 80% acetonitrile <!> (HPLC grade), 20% isopropanol <!> (see Step 9).

Equipment

Commercial heat gun (Weller 6966) or an oven set at 200°C (see Step 4)
100 μm ID x 365 μm OD Fused silica capillary (FSC) tubing
200 μm ID x 365 μm OD Fused silica capillary tubing
Pressurized loading vessel (see Experiment 4)

PROCEDURE

IMAC Column Construction

1. Cut a 20-cm long piece of fused silica capillary (200 μm ID x 360 μm OD).

2. Add 17 μL of formamide to 88 μL of Kasil #1 to prepare a Kasil slurry. After adding the formamide, vortex and centrifuge the slurry.

   It is important to add the formamide to the Kasil and not the reverse to ensure proper polymerization of the mixture. Kasil #1 is a potassium silicate solution. The addition of formamide causes Kasil to polymerize. It forms a porous frit at the end of the FSC column, where it retains the chromatographic stationary phase but allows the mobile phase solutions and solubilized peptides to flow through the column.

3. Briefly touch the FSC column to the top of the Kasil slurry (Fig. 2).

   The Kasil solution will be pulled into the column through capillary action. You should end up with an ~0.5-cm long Kasil frit.

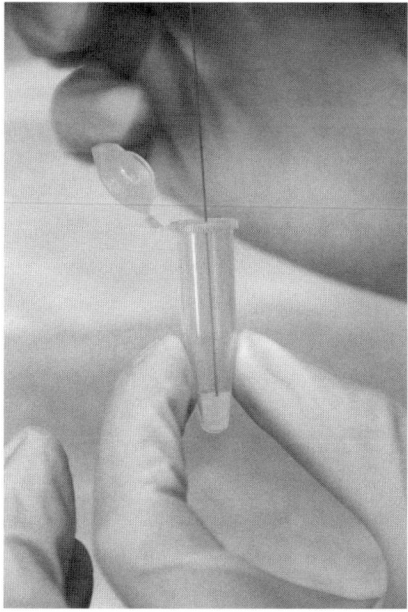

FIGURE 2. Creating a Kasil frit by touching the surface of the FSC tubing into a Kasil slurry.

88 EXPERIMENT 5

4. Heat the column for several seconds with a commercial heat gun (Fig. 3). The Kasil frit will turn white. Alternatively, heat the column for 3 minutes in an oven at 200°C.

5. Place the fritted column into the pressurized loading vessel. Rinse the column with acetonitrile to check for flow and to remove excess Kasil (Fig. 4).

   At a pressure of 200 psi, the column should flow at a steady rate of >10 µL/min.

6. Pack the fritted column with a slurry of POROS MC 20 IMAC chromatography resin in H_2O to a height of 10 cm.

   After packing the column, an optional back frit can be made at the open end of the column by repeating Steps 3 and 4. With a back frit, the mobile phase flow through the column can be reversed. This can improve column washing (in Protocol 4).

Reversed-phase Capture Column (Precolumn) Construction

7. Cut a 20-cm-long piece of fused silica capillary (100 µm ID × 360 µm OD).

   The precolumn will be used to capture phosphopeptides released from the IMAC column in Protocol 4 of this experiment. The captured peptides will be transferred onto an analytical microcapillary RP column for LC-MS/MS analysis.

8. Repeat Steps 2–5 with the precolumn FSC to make a Kasil frit at one end of the FSC.

9. Using the pneumatic loading vessel, pack the precolumn with a slurry of YMC-Gel ODS-A 12 nm S-10 µm reversed phase medium in 80% acetonitrile, 20% isopropanol.

   Typical packing length is 10 cm. No back frit is needed.

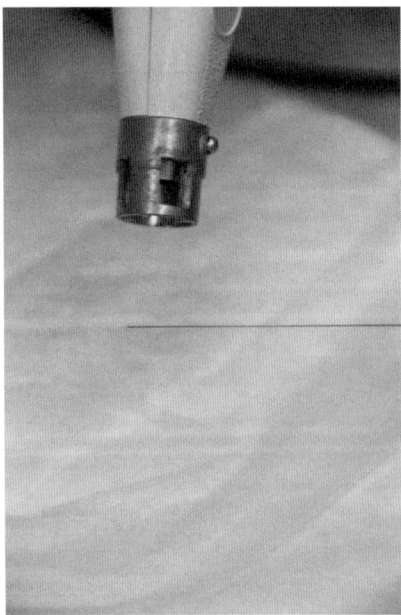

FIGURE 3. Using a heat gun to bake the Kasil to make a porous frit for making FSC microcapillary HPLC columns.

FIGURE 4. Using the loading bomb to wash and pack the fritted FSC HPLC column with HPLC mobile phases and the IMAC resin.

Optional Kasil frit-free column construction

Creating Kasil frits can be cumbersome, and variations in frit construction can lead to irreproducible chromatography. Alternatively, frit columns with the Upchurch M-120x in-line microfilter in the manner outlined in the MudPIT column construction section of this manual (Experiment 6). This will also remove the need to use Teflon tubing to attach the IMAC and reversed-phase capture columns together—a process that often damages one or both of the columns.

PROTOCOL 3

Preparation of the IMAC Column

MATERIALS

CAUTION: See Appendix 11 for appropriate handling of materials marked with <!>.

Reagents

0.1% Acetic acid (99.9%+) <!>
100 mM EDTA (pH 8.5)
100 mM Fe(III) chloride (Sigma-Aldrich 451649) <!>

Equipment

IMAC column (from Protocol 2)
Pressurized loading vessel

PROCEDURE

Preprocessing the IMAC column

1. Place the IMAC column into the pressurized loading vessel (see Experiment 4, Protocol 2).
2. Rinse the IMAC column with 100 mM EDTA (pH 8.5) for at least 10 minutes at 10–12 μL/min.
3. Rinse the IMAC column with MilliQ-H_2O for at least 10 minutes at 10–12 μL/min.

Conditioning the IMAC column

4. Rinse the IMAC column with 100 mM Fe(III) chloride for 10–15 minutes at 10–12 μL/min.
5. Rinse the IMAC column with 100 mM Fe(III) chloride for 10–15 minutes at 2–5 μL/min in the reverse direction.

    The step can only be performed if the column contains a back frit (see Step 6 of Protocol 2).

6. Flip the column back and rinse the IMAC column with 0.1% acetic acid for 10 minutes at 10–12 μL/min.

PROTOCOL 4

Processing the Sample

MATERIALS

CAUTION: See Appendix 11 for appropriate handling of materials marked with <!>.

Reagents

0.1% Acetic acid <!>
Elution buffer (250 mM sodium phosphate; Sigma-Aldrich S-9390)
Lyophilized, esterified peptide sample in glass vial (Protocol 1, Step 20)
Organic rinse solution (25% acetonitrile, 100 mM NaCl, 1% acetic acid)
Rehydration solution (30 µL methanol <!>/30 µL acetonitrile <!>/30 µL 0.1% acetic acid)
Rinse solution (0.3 M guanidine-HCl <!> in 95% ethanol <!>)
Solution A (0.2 M acetic acid in H_2O)
Solution B (0.2 M acetic acid in 70% acetonitrile)

Equipment

IMAC column, conditioned (from Protocol 3)
Tandem mass spectrometer with nanoESI source (see Experiment 4)
Pressurized loading vessel (i.e., the bomb)
Reversed-phase precolumn (from Protocol 2)
0.012 inches ID x 0.060 inches OD Teflon tubing
Vial, glass

PROCEDURE

Loading the Sample

1. Resuspend the lyophilized peptides in 90 µL of rehydration solution.
2. Place a conditioned IMAC column into the pressurized loading vessel along with a glass vial of 0.1% acetic acid solution. Determine the pressure needed to establish a flow rate of 1 µL/min.
3. Equilibrate the IMAC column with 0.1% acetic acid solution for 10 minutes at 1 µL/min.
4. Replace the acetic acid with the sample vial.
5. Load the sample onto the IMAC column at no more than 1 µL/min. Collect the flowthrough fraction, which should contain nonphosphorylated peptides.

Washing the IMAC Column for Elution

6. Rinse the IMAC column with 0.1% acetic acid for 10 minutes at 1–2 µL/min.

    This step is used to push the sample onto the column.

7. Rinse with the organic rinse solution for 5 minutes at 10 µL/min.
8. If the IMAC column has a back frit, flip the column assembly over and rinse for 5 minutes in the reverse direction at 10 µL/min.
9. Rinse with 0.1% acetic acid for 10 minutes at 10 µL/min.

Eluting and Analyzing the Phosphopeptides

10. Using a piece of Teflon tubing (0.012 inches ID x 0.060 inches OD), attach the RP-precolumn to the IMAC column (Fig. 5A–D). Test the connection with 0.1% acetic acid at 600 psi.
11. Replace the acetic acid with elution buffer.
12. Elute the phosphopeptides from the IMAC column onto the precolumn with 40 µL of elution buffer.

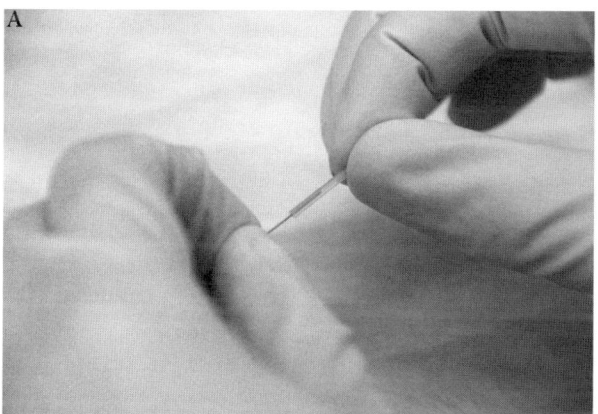
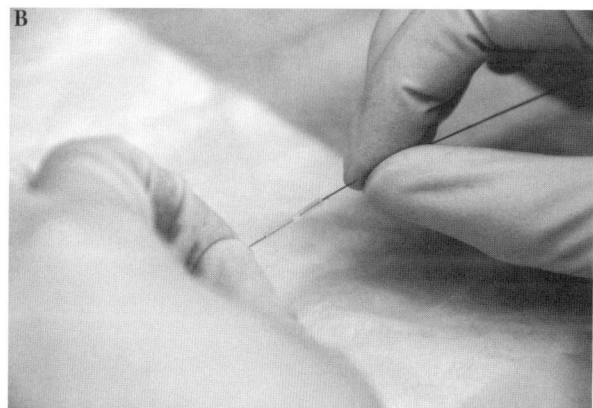
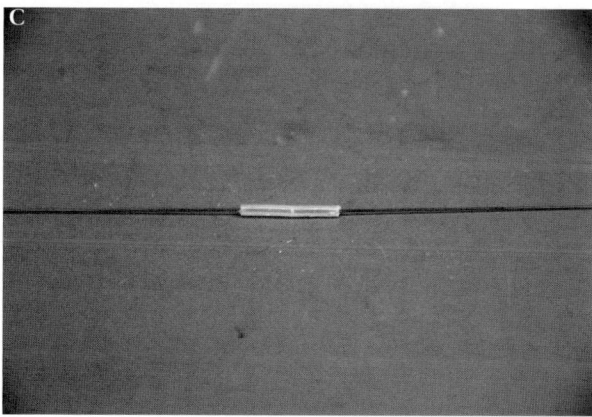
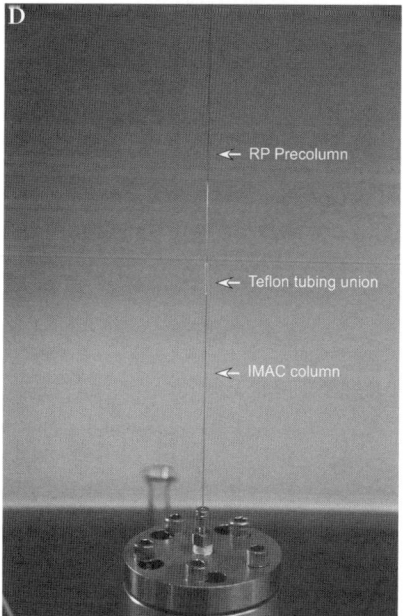

FIGURE 5. (A) Attaching a piece of Teflon tubing to the IMAC column for connecting the column to the precolumn. (B) Attaching the precolumn to the IMAC column. (C) Close-up view of the IMAC and precolumn joined together using a piece of Teflon tubing. (D) Using the loading bomb to transfer phosphopeptides from the IMAC column to the precolumn.

13. Transfer the precolumn with the captured phosphopeptides to the mass spectrometry instrument (Fig. 6A). Flush the precolumn with 0.2 M acetic acid in H$_2$O for 10 minutes.

    Washing the precolumn with the 0.2 M acetic acid solution removes the sodium phosphate in the elution buffer, which suppresses ionization during ESI.

14. Using a piece of Teflon tubing (0.012 inches ID x 0.060 inches OD), attach the precolumn to a pulled microcapillary RP HPLC column (Fig. 6B and Experiment 4, Protocol 2). As described in Experiment 4, perform LC-MS/MS analysis on the sample.

    The White lab uses 0.2 M acetic acid in H$_2$O for buffer A and 0.2 M acetic acid in 70% acetonitrile for buffer B to run the RP-HPLC gradient.

15. Search and analyze the acquired MS/MS data against a protein database using the methods described in Experiment 8.

    Search parameters should include differential modification of +80 Da (presence or absence of phosphate) on serine, threonine, and tyrosine and a static modification of +14 Da (methyl groups) on aspartic acid, glutamic acid, and the carboxyl terminus.

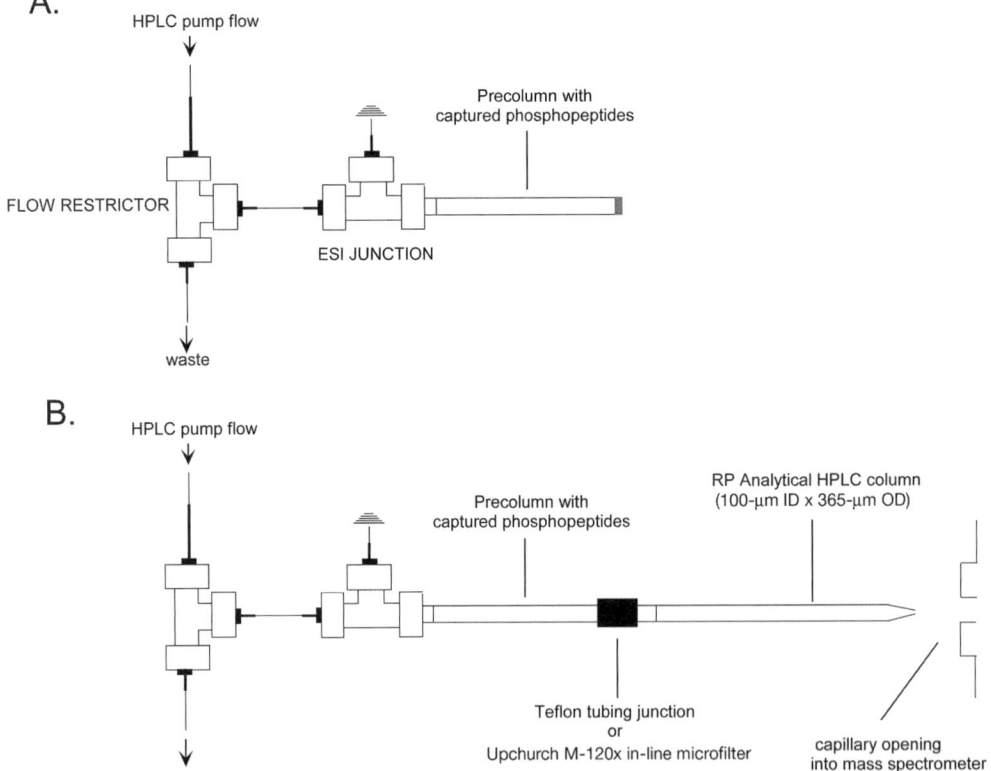

FIGURE 6. Diagram of the FSC connections used to wash and analyze the captured phosphopeptides. (A) Precolumn with the capture phosphopeptides connected to the mass spectrometer's LC system for washing the column with mobile phase buffer A. (B) RP microcapillary analytical HPLC column coupled to the precolumn for LC-MS/MS of the captured phosphopeptides (Experiment 4). The RP precolumn is coupled to the analytical RP column using either Teflon tubing or an Upchurch union.

REFERENCES

Andersson L. and Porath J. 1986. Isolation of phosphoproteins by immobilized metal (Fe^{3+}) affinity chromatography. *Anal. Biochem.* **154:** 250–254.

Beausoleil S.A., Jedrychowski M., Schwartz D., Elias J.E., Villén J., Li J., Cohn M.A., Cantley L.C., and Gygi S.P. 2004. Large-scale characterization of HeLa cell nuclear phosphoproteins. *Proc. Natl. Acad. Sci.* **101:** 12130–12135.

Brill L.M., Salomon A.R., Ficarro S.B., Mukherji M., Stettler-Gill M., and Peters E.C. 2004. Robust phosphoproteomic profiling of tyrosine phosphorylation sites from human T cells using immobilized metal affinity chromatography and tandem mass spectrometry. *Anal. Chem.* **76:** 2763–2772.

Ficarro S.B., McCleland M.L., Stukenberg P.T., Burke D.J., Ross M.M., Shabanowitz J., Hunt D.F., and White F.M. 2002. Phosphoproteome analysis by mass spectrometry and its application to *Saccharomyces cerevisiae*. *Nat. Biotechnol.* **20:** 301–305.

Ficarro S.B., Salomon A.R., Brill L.M., Mason D.E., Stettler-Gill M., Brock A., and Peters E.C. 2005. Automated immobilized metal affinity chromatography/nano-liquid chromatography/electrospray ionization mass spectrometry platform for profiling protein phosphorylation sites. *Rapid Commun. Mass Spectrom.* **19:** 57–71.

Kim J.E., Tannenbaum S.R., and White F.M. 2005. Global phosphoproteome of HT-29 human colon adenocarcinoma cells. *J. Proteome Res.* **4:** 1339–1346.

Larsen M.R., Thingholm T.E., Jensen O.N., Roepstorff P., and Jørgensen T.J. 2005. Highly selective enrichment of phosphorylated peptides from peptide mixtures using titanium dioxide microcolumns. *Mol. Cell Proteomics* **4:** 873–886.

Mann M., Ong S.E., Grønborg M., Steen H., Jensen O.N., and Pandey A. 2002. Analysis of protein phosphorylation using mass spectrometry: Deciphering the phosphoproteome. *Trends Biotechnol.* **20:** 261–268.

Moser K. and White F.M. 2006. Phosphoproteomic analysis of rat liver by high capacity IMAC and LC-MS/MS. *J. Proteome Res.* **5:** 98–104.

Pinkse M.W., Uitto P.M., Hilhorst M.J., Ooms B., and Heck A.J. 2004. Selective isolation at the femtomole level of phosphopeptides from proteolytic digests using 2D-NanoLC-ESI-MS/MS and titanium oxide precolumns. *Anal. Chem.* **76:** 3935–3943.

Posewitz M.C. and Tempst P. 1999. Immobilized gallium(III) affinity chromatography of phosphopeptides. *Anal. Chem.* **71:** 2883–2892.

Rush J., Moritz A., Lee K.A., Guo A., Goss V.L., Spek E.J., Zhang H., Zha X.M., Polakiewicz R.D., and Comb M.J. 2005. Immunoaffinity profiling of tyrosine phosphorylation in cancer cells. *Nat. Biotechnol.* **23:** 94–101.

EXPERIMENT 6

Multidimensional Protein Identification Technology (MudPIT) Analysis of Whole-cell Lysates

OVERVIEW OF MULTIDIMENSIONAL SEPARATION AND IDENTIFICATION OF PROTEINS

Rapid and accurate identification and quantification of proteins and their posttranslational modifications, either globally or in the context of defined protein complexes, is one of the challenges of proteomics. The development of proteomic technologies to profile purified protein complexes or the entire proteome has taken on an increased sense of urgency now that genome sequencing projects have provided the sequences for all the possible proteins in many organisms. A key part of this advancement will rely on the separation sciences to resolve the individual components of complex protein and peptide mixtures prior to analytical analysis. Multidimensional separations are the solution when a separation based on any single parameter does not provide the required resolution and capacity. In the past, multidimensional separations have been used in profiling the components of complex mixtures or purifying specific target proteins or peptides.

Multidimensional separation typically relies on using two or more independent physical properties of the peptides to fractionate the mixture into individual components (Fig. 1). When this is the case, the separation methods are considered "orthogonal." Physical properties commonly exploited include size, charge, hydrophobicity, and biological interaction or affinity. Typically, components that are not separated in the first separation step are resolved in the second. Peak capacity is the number of individual components that can be resolved by a separation method. Giddings (1987) provides a mathematical model showing that, if the multidimensional separations are orthogonal, then the total peak capacity is the product of the individual peak capacities of each separation. Load capacity is defined as the maximum amount of material that can be run in a separation while maintaining chromatographic resolution. Multidimensional separations can be used to increase the load capacity for analyzing low abundance or trace components in a peptide mixture. A multidimensional system is considered "comprehensive" when the entire eluent from the first dimension separation is fractionated on the second dimension.

For multidimensional separations of peptides using ion-exchange chromatography, cation-exchange chromatography has been the method of choice for the first dimension. At a pH<3, the negative charges at carboxyl groups and the carboxyl terminus are neutralized due to the complete protonation, leaving arginine, lysine, and histidine residues, plus the amino terminus, to contribute to a net positive charge of the peptide. The fully protonated peptides can be fractionated by strong cation exchange (SCX) chromatography. Although the major property governing peptide retention during ion exchange (IEX) involves ionic interactions, most ion exchangers also exert

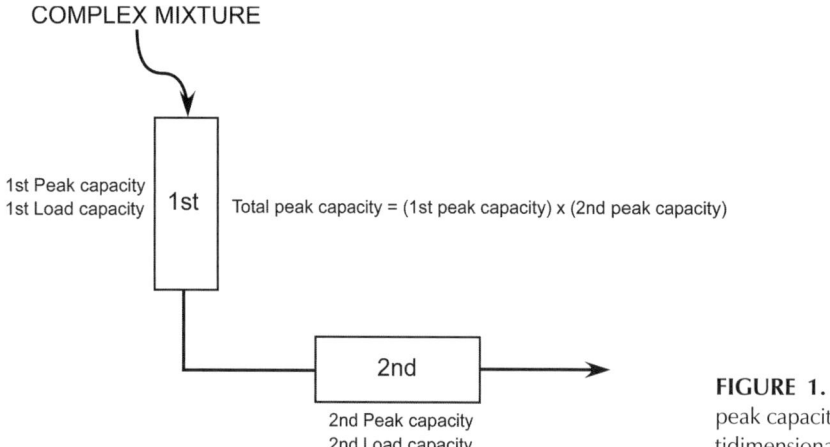

FIGURE 1. Diagram of increased load and peak capacity achieved using orthogonal multidimensional separations.

some hydrophobic influence. This is commonly referred to as a mixed-mode effect. The mixed-mode effect partially explains why peptides with the same number of positively charged residues can be resolved by IEX.

Methods using liquid chromatography coupled with tandem mass spectrometry allow direct identification of proteins in mixtures. We developed a rapid, automated, and sensitive process for comprehensively resolving and identifying large numbers of proteins in complex mixtures originally published under the name Direct Analysis of Large Protein Complexes (DALPC) (Link et al. 1999). It is now published as Multidimensional Protein Identification Technology (MudPIT) (Washburn et al. 2001). This powerful technique uses multidimensional liquid chromatography and tandem mass spectrometry to separate and fragment peptides (Fig. 2). Computational com-

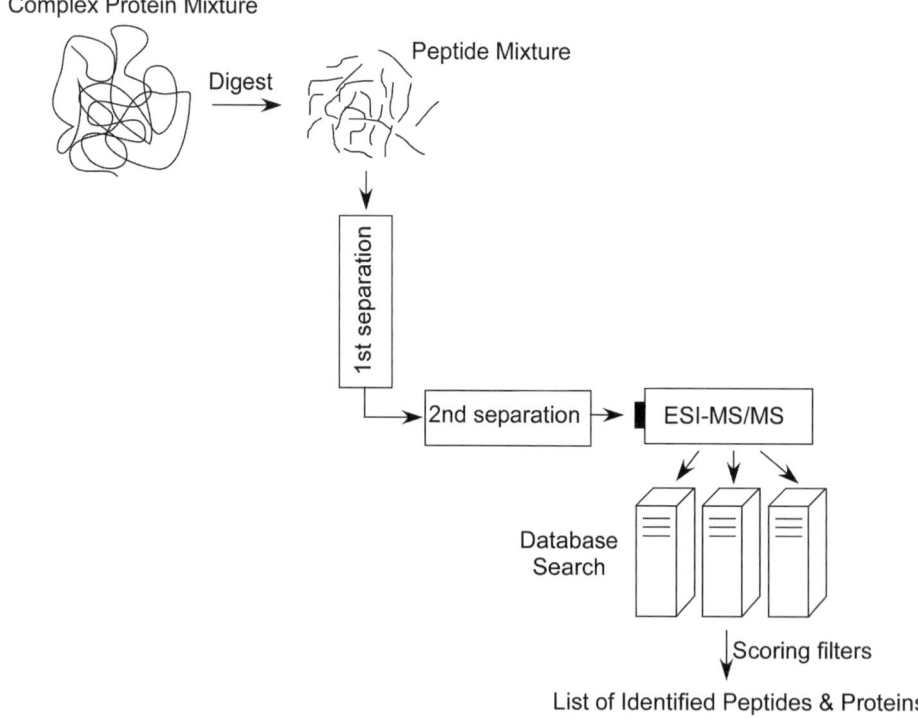

FIGURE 2. Overview of MudPIT analysis to identify proteins in complex protein mixtures.

parison of tandem spectra with genomic sequence databases and data processing produces a list of proteins in the starting sample. Two-dimensional separation can be performed in a single biphasic column, which has improved resolution and loading capacity compared to a single-dimension column (Link et al. 1999).

MudPIT avoids several of the drawbacks inherent when 1D or 2D gel electrophoresis and mass spectrometry are coupled for the separation and identification of proteins in complex mixtures. These limitations include limited fractionation ranges, protein insolubility, and limited recovery of material. Careful analysis of published studies shows that only the most abundant bands or spots are typically selected from gels for identification by mass spectrometry, and faint bands or spots are routinely neglected. Multidimensional chromatography combined with sophisticated data-dependent mass spectrometry data acquisition methods (i.e., dynamic exclusion, etc.) have increased the dynamic range of MudPIT for detecting low-abundance proteins in complex mixtures. As mass spectrometry has the sensitivity to identify proteins in quantities undetectable by silver staining, direct analysis of protein complexes enables these rare proteins to be identified.

MULTIDIMENSIONAL PROTEIN IDENTIFICATION TECHNOLOGY

In MudPIT, a digested protein complex is loaded directly onto a biphasic SCX-RP or triphasic RP-SCX-RP microcapillary column that directly elutes into an ESI mass spectrometer (Fig. 3). For the biphasic column, the peptide sample is desalted off-line before loading onto the MudPIT column. The triphasic column is used to desalt the sample online to minimize loss of material. The peptides are sequentially eluted from the capillary column and fragmented in the mass spectrometer. In this protocol, MudPIT is carried out using a ThermoFisher LTQ ion trap mass spectrometer and an Agilent quaternary HPLC pump. The biphasic or triphasic MudPIT column is interfaced directly to the mass spectrometer and HPLC pump using a fused silica capillary nanoESI assembly (see Fig. 5 on p. xxx). The HPLC pump is directly controlled by the LTQ's Xcalibur software. Within the instrument setup of Xcalibur, the gradient of each individual step is programmed (Appendix 7). A typical analysis consists of a fully automated 6-step chromatography run on highly complex mixtures, but any number of steps can be applied to the column.

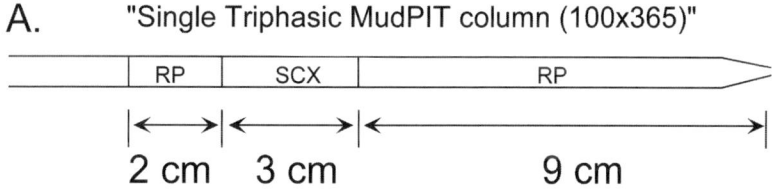

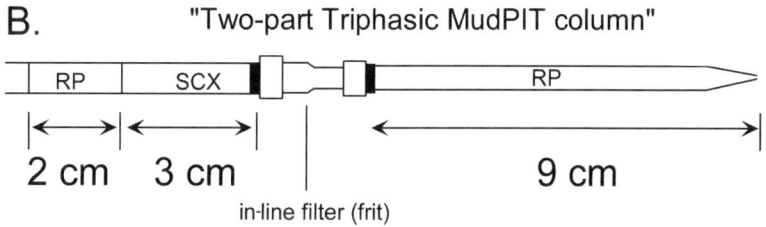

FIGURE 3. Schematic of two types of triphasic MudPIT columns. (A) All three phases packed into a single, fritted microcapillary FSC HPLC column. (B) A mechanical frit is used to create two separate columns joined using a zero-dead volume union.

CONSTRUCTION OF MudPIT MICROCAPILLARY HPLC COLUMNS

There are two ways of constructing MudPIT assemblies. In the first approach, a biphasic or triphasic MudPIT column is packed in a single 100 x 365 µm pulled microcapillary column (see Fig. 3A). In a second approach, a MudPIT assembly is constructed in two parts and connected using a zero dead-volume union (Fig. 3B). The first part is an SCX or RP-SCX column constructed using a microfilter as a mechanical frit. A pulled RP column is connected to the SCX or RP-SCX column using a zero dead-volume union to create the biphasic or triphasic column. Because it is easier to pack and load MudPIT columns using this second approach, we use this strategy here (Link et al. 2003).

PROTOCOL 1

Analysis of Whole-cell Lysates Using MudPIT

MATERIALS

CAUTION: See Appendix 11 for appropriate handling of materials marked with <!>.

Reagents

5% Acetonitrile <!>, 0.1% formic acid <!>
    Use HPLC-grade acetonitrile.
Desalting buffer A (2% acetonitrile, 0.1% TFA <!>)
Desalting buffer B (95% acetonitrile, 0.1% TFA)
Desalting buffer C (70% formic acid, 30% isopropanol <!>)
MudPIT buffer A (5% acetonitrile, 0.1% formic acid)
MudPIT buffer B (80% acetonitrile, 0.1% formic acid)
MudPIT buffer C (500 mM ammonium acetate, 5% acetonitrile, 0.1% formic acid)
Peptide sample
Reversed-phase resin (Phenomenex Synergi 4u Hydro-RP 80A)
Strong cation exchange (SCX) chromatography resin (Whatman 5 µm PartiSphere SCX)

Equipment

Calibrated glass pipette, 5 µL
Fused capillary cleaving scribe (New Objective)
Fused silica capillary tubing, 100 µm ID x 365 µm OD (PolyMicro Technologies)
Fused silica capillary tubing, 75 µm ID x 365 µm OD
Fused silica capillary tubing, 50 µm ID x 365 µm OD
HPLC quaternary pump (Agilent)
In-line microfilter (M-120x; Upchurch)
Mass spectrometer, LCQ or LTQ ion trap (ThermoFisher)
Microfingertight fitting (F-125; Upchurch)
nanoESI assembly (see Experiment 4 and Appendix 1)
PEEK MicroTight adapters

PEEK 380 μm ID MicroTight sleeve
PEEK MicroTight ZDV union (P704; Upchurch)
Pressurized loading vessel
Sample loop, 10 μL or larger
Syringe, 100-μL
Syringe port

PROCEDURE

Packing a Multidimensional Column

1. Cut ~7 inches of 100 μm ID × 365 μm OD FSC with a fused capillary cleaving scribe.

   When analyzing large amounts of material (e.g., mg's), use 200 μm ID × 365 μm OD FSC. This larger inner diameter of FSC is very fragile and should be handled with extreme care.

2. Insert the FSC into the green sleeve of an Upchurch F-125 microfingertight fitting. Screw an Upchurch M-120x in-line microfilter end fitting over the capillary assembly. This will act as a mechanical frit.

3. Place the FSC into the pressurized loading vessel with the SCX packing material. Pack 3 cm of SCX.

4. Slowly release the pressure and replace the SCX slurry with 5% acetonitrile, 0.1% formic acid. Wash the column for several minutes.

5. For triphasic columns, replace the 5% acetonitrile, 0.1% formic acid with RP packing material. Pack 2 cm of RP. Equilibrate the column for several minutes with 5% acetonitrile, 0.1% formic acid.

Loading Samples onto a Multidimensional Column

6. Connect a syringe port to a 10-μL or larger sample loop using a PEEK ZDV union (Fig. 4A). Connect the opposite end of the loop to a piece of 75 × 365 μm FSC using a PEEK MicroTight ZDV adapter and a PEEK sleeve.

   The 10-μL sample loop can be replaced with a 50-μL or 100-μL loop, depending on the sample volume. The use of a sample loop and the HPLC pump allows higher pressure to quickly and reliably load peptide samples onto the MudPIT columns.

7. Use a 100-μL syringe to clean and equilibrate the sample loop. Wash 2 times with desalting buffer B, 2 times with desalting buffer C, and 2 times with desalting buffer A.

   The desalting buffers A, B, and C are used to remove any trace peptide material from earlier experiments and to fill the loop with the acqueous mobile phase buffer A.

8. Inject 10–100 μL of the peptide sample into the sample loop via the syringe port. Remove the port and replace with a MicroTight adapter. Connect the sample loop to the biphasic column using the assembly shown (Fig. 4B).

9. The MudPIT loading assembly is constructed as described (see Fig. 4B).

   a. Connect the typical 1/16-inch transfer line from the HPLC pump (intakes connected to MudPIT buffers A, B, and C) to a 75 μm ID × 365 μm OD 10-cm long FSC transfer line using a PEEK MicroTight adapter and a PEEK 380 μm ID MicroTight sleeve.

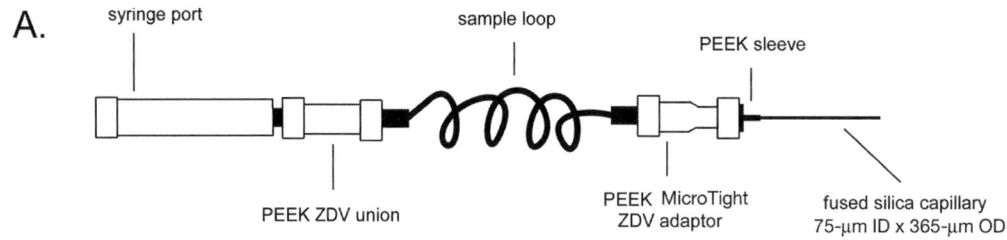

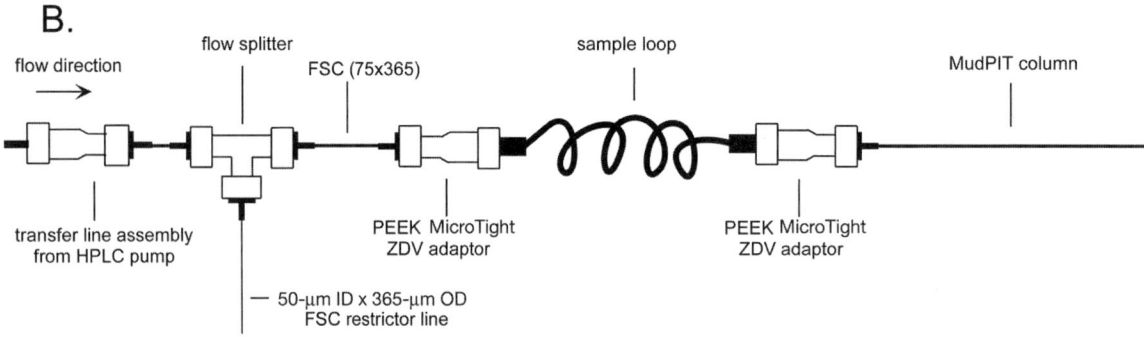

FIGURE 4. Schematic of connections for off-line loading of samples onto MudPIT columns using sample loops. (A) Loop assembly for loading samples into a loading loop. (B) Loading assembly for transferring sample from the loading loop onto the MudPIT column.

    b. Connect the 75 μm ID × 365 μm OD FSC transfer line from the PEEK MicroTight adapter to a PEEK MicroTee using a PEEK 380 μm ID MicroTight sleeve.

    c. Connect a PEEK MicroTight adapter on the opposite arm using a PEEK MicroTight ZDV union, a 5-cm piece of 75 × 365 μm FSC, and PEEK MicroTight sleeves.

    d. Connect the sample loop to the PEEK MicroTight adapter.

    e. Connect the MudPIT column to a sample loop using a PEEK MicroTight adapter and a PEEK 380 μm ID MicroTight sleeve.

    f. Prepare a restrictor line by connecting a 30-cm piece of 50 μm ID × 365 μm OD FSC. Connect the FSC through the third arm of the Tee using a PEEK MicroTight sleeve.

> The MicroTee splits the flow between the MudPIT column and the restrictor line. The HPLC pump is set at a constant flow rate and the length of the restrictor line is adjusted to deliver the desired flow rate to the HPLC column. This splitter assembly allows the HPLC pump to be operated at flow rates that are delivered accurately. Typically, the pump is set at 300 μL/min and the restrictor line is shortened to give a 300-to-1 flow split for an effective 1 μL/min flow rate to the HPLC column. The flow through the restrictor line is collected in a waste bottle and discarded.

10. Set the HPLC pump to 100% MudPIT buffer A with a flow rate of 300 μL/min. The HPLC pump flow rate is held constant throughout the loading process.

11. Once you have equilibrated the column, trim the 50 × 365 μm restrictor line using the quartz capillary cleaving tool until a flow rate of 0.5 μL/min through the MudPIT column is reached as measured for 1 minute using a 5-μL calibrated glass pipette.

> Measurement of the flow rate and adjustment of the split line may have to be repeated several times until the target flow rate is reached.

12. Run 100% MudPIT Buffer A until you have displaced the sample from the sample loop onto the MudPIT column.

13. Disconnect the sample loop and reconnect the biphasic or triphasic column to the pump.
14. Connect a pulled RP column to the biphasic column with an in-line microfilter assembly.
15. Equilibrate the column for 10 minutes with 100% MudPIT Buffer A.

Preparing a Multidimensional Column for Electrospray Ionization Mass Spectrometry

16. After equilibration of the column, it may be necessary to cut the column due to its long length.
    a. Remove the pulled RP column from the microfilter assembly.
    b. Using a cleaving tool, carefully cut the FSC 2 cm past the end of the packed material.
    c. Reconnect the RP column to the microfilter assembly and cut the FSC 2 cm past the end of the biphasic packed material.

    > Depending on your source, it may be necessary to slide a MicroTight union through the back end of the RP column. This will hold the column in-line with the opening of the mass spectrometer.

17. Install the MudPIT column in the MudPIT ESI assembly (see Fig. 5). Set the HPLC pump to 100% MudPIT buffer A with a flow rate of 200 µL/min. Once you have equilibrated the column, trim the 50 × 365 µm restrictor line using the quartz capillary cleaving tool until a flow rate of 0.5 µL/min through the MudPIT column is reached as measured for 1 minute using a 5-µL calibrated glass pipette. Carefully position the pulled microcapillary HPLC column tip at the entrance of the mass spectrometer (see Experiment 4).

    > Measurement of the flow rate and adjustment of the split line may have to be repeated several times until the target flow rate is reached.

18. Using Xcalibur software, program the HPLC gradient and mass spectrometer instrument files for MudPIT operation (see Appendix 7).

19. START!!!

    > Upon completion of the run, process and analyze the acquired tandem mass spectrometry data files as described in Experiment 8.

    > This 6-step MudPIT run takes approximately 9 hours to complete.

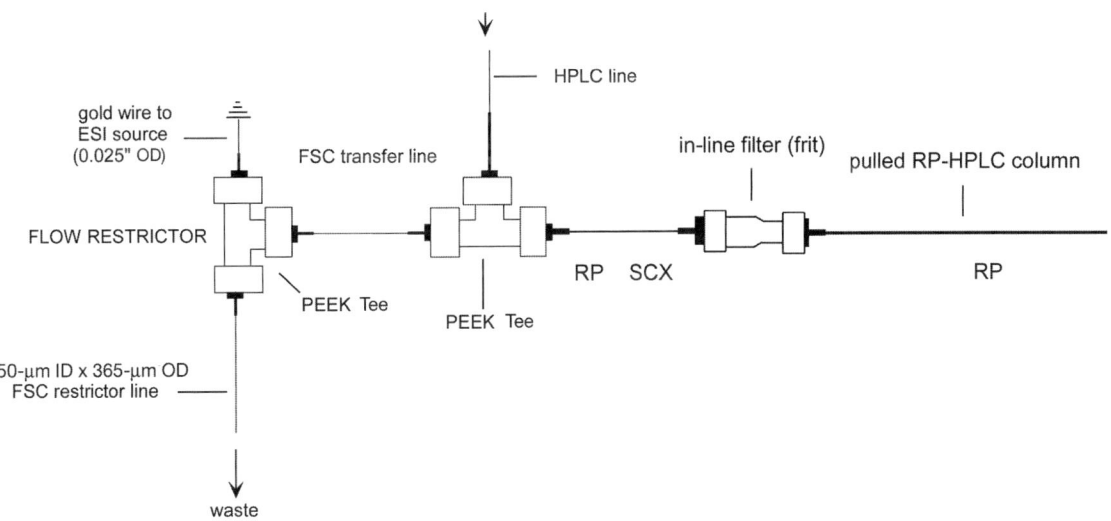

FIGURE 5. Schematic of the FSC connections for performing MudPIT using a nanoESI source.

REFERENCES

Giddings J.C. 1987. Concepts and comparisons in multidimensional separation. *J. High Resolut. Chromatogr. Chromatogr. Commun.* **10**: 319–323.

Link A.J., Jennings J.L., and Washburn M.P. 2003. Analysis of protein composition using multidimensional chromatography and mass spectrometry. In *Current protocols in protein science* (ed. J.E. Coligan et al.), chapter 23, pp. 1–25. John Wiley and Sons, New York.

Link A.J., Eng J., Schieltz D.M., Carmack E., Mize G.J., Morris D.R., Garvik B.M., and Yates J.R. III. 1999. Direct analysis of protein complexes using mass spectrometry. *Nat. Biotechnol.* **17**: 676–682.

Washburn M.P., Wolters D., and Yates J.R. III. 2001. Large-scale analysis of the yeast proteome by multidimensional protein identification technology. *Nat. Biotechnol.* **19**: 242–247.

EXPERIMENT 7

Quantitative Mass Spectrometry Analysis of Whole-cell Extracts (iTRAQ)

Eric S. Simon

*Department of Biological Chemistry, University of Michigan,
Ann Arbor, Michigan 48109*

The field of proteomics, in its brief history, has progressed rapidly from merely identifying proteins within large, complex mixtures to additionally determining the relative abundance of proteins between two or more samples. This progression to quantitative measurements can be attributed to the advancements made in mass spectrometry and the adoption of stable isotope labeling strategies for distinguishing the relative abundance of like peptides between samples (Gygi et al. 1999; Ong and Mann 2005). Typically, a set of reagents consists of one tag that has a natural isotope distribution and a complementary tag consisting of the exact elemental composition and chemical structure, except that it is isotope enriched, typically with ^{13}C, although ^{2}H and ^{15}N can be used as well. The isotope-enriched tag, or heavy reagent, is chemically equivalent to the light reagent but has a higher mass. In a typical workflow, peptides from a control sample are labeled with the light reagent, and an experimental sample is labeled with the heavy reagent. The two labeled peptide samples are then mixed, fractionated, and analyzed by mass spectrometry. A peptide present in both samples will have essentially the same physicochemical properties, including charge state, isoelectric point, and hydrophobicity, despite having tags of slightly different mass. As a peptide elutes, relative quantification is determined in the MS mode by integrating the peaks under the extracted ion chromatograms of each of the light and heavy labeled peptides across their chromatographic peak (Ong and Mann 2005). Figure 1 provides an example of this concept. Examples of stable isotope labeling strategies that yield MS-mode quantification in this way are stable isotope labeling by amino acids in cell culture (SILAC) (Ong et al. 2002) and isotope-coded affinity tag (ICAT) (Gygi et al. 1999).

Despite the effectiveness of these strategies, no more than two or three samples can be compared in one experiment. This is due to the inherent increase in complexity of the spectra in the MS mode when multiple isotope-enriched tags are used. For example, if a control is tagged with a light reagent and the test sample is tagged with a corresponding, isotope-enriched heavy reagent, then the mass spectrometer registers two peaks representing that peptide as opposed to just one. As the number of samples increases, the number of higher mass reagents increases, thus increasing the number of peaks in the mass spectrum. This envelope of peaks representing the same peptide can overlap in the spectrum with envelopes of other peptides and can be extremely difficult to sort out.

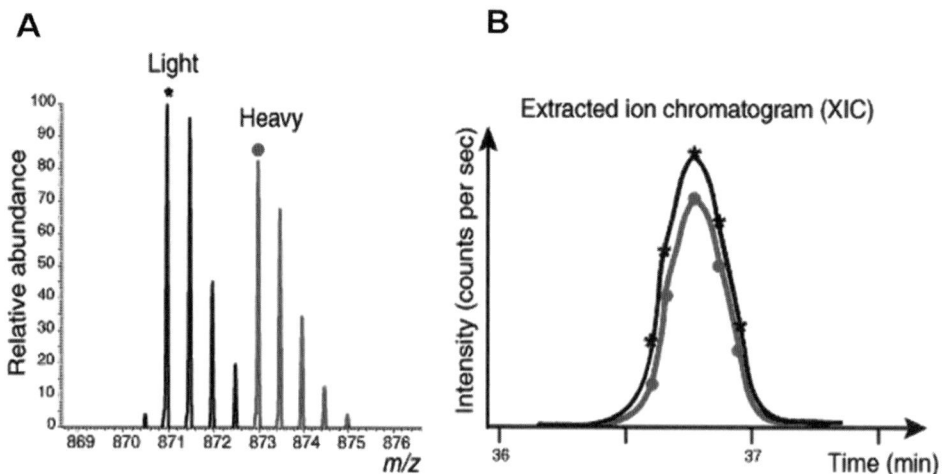

FIGURE 1. (A) Isotope clusters of a peptide labeled with light and heavy stable isotope reagents for relative quantification. The intensity of the monoisotopic peak of the light (*) and heavy (•) analog across the peptides elution profile is called the extracted ion chromatogram for each ion (XIC) (B). The area under each XIC correlates to the relative abundance of the peptide between the two samples. (Figure adapted from Ong and Mann 2005.)

INTRODUCTION TO iTRAQ

Recently, a commercial reagent has emerged that makes it possible to analyze multiple samples simultaneously. Known as the isobaric tag for relative and absolute quantification (iTRAQ), the original reagent, which is used in the CSHL Proteomics course, can compare up to four samples in parallel (Ross et al. 2004), while a more current version can handle multiplexing of up to eight samples (Choe et al. 2007). The ability of iTRAQ to compare relative protein abundances across as many as eight samples is a significant advantage over other stable isotope strategies, like SILAC and ICAT.

Figure 2 provides an illustration of the iTRAQ concept. The reagent itself is made up of three components. The reporter group, an N-methylpiperazine moiety, serves as the quantitative component, or reporter ion, observed in the low mass region of MS/MS spectra. Depending on the tag used, it will have a mass of 114, 115, 116, or 117 Da. The next component of iTRAQ is the mass balancer, a carbonyl group, which is isotope enriched to have an offset mass of 31, 30, 29, or 28 Da, depending on the mass of the reporter group it is paired with. The reporter and mass balancer combine to yield a mass equivalence of 145 Da [(114 + 31), (115 + 30), (116 + 29), and (117 + 28)] across a set of four isobaric tags. The third component of iTRAQ is an amine-reactive group, N-hydroxysuccinimide, which derivatizes peptides at primary amines, thus targeting the amino termini and lysine side chains. Once the peptides have been derivatized, they are mixed together. A peptide present in all four samples will appear as a single peak in an MS survey scan. However, once fragmented, the peptide's MS/MS spectrum will contain the signature series of reporter peaks at 114–117 Da that are used for quantification. Also present in the MS/MS spectrum are a series of peaks representing amide backbone fragments, like b and y ions, that are used for peptide identification. Figure 3 displays two spectra acquired on a MALDI TOF/TOF mass spectrometer that illustrate these concepts. Figure 3A shows a precursor spectrum of iTRAQ-labeled peptides. Figure 3B shows an MS/MS spectrum of the precursor ion of *m/z* 1626.0. Each of the iTRAQ tags contains an isotope distribution such that the derivatized peptides are indistinguishable in MS spectra but yield signature ions in the low mass region of MS/MS spectra. The inset (Fig. 3B) shows the region

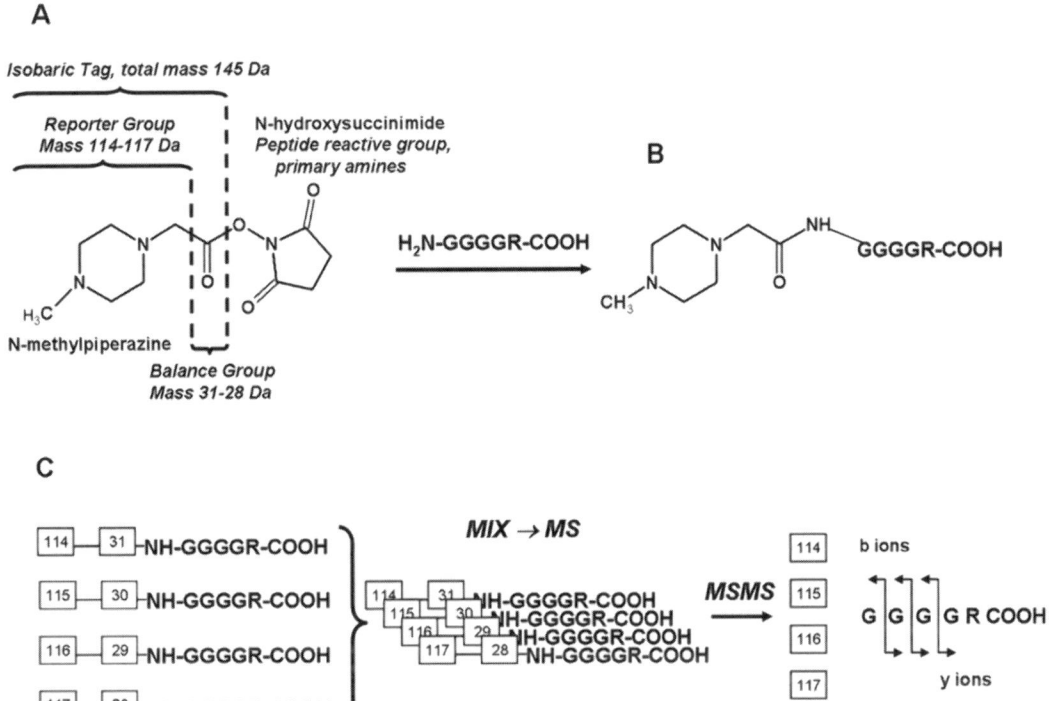

FIGURE 2. (A) The functional components of the iTRAQ reagent. (B) The N-hydroxysuccinimide component reacts with primary amines, resulting in modification of peptides on the amino-terminus and side-chain lysine residues. (C) After mixing peptides with different tags, they are indistinguishable in MS mode, but subsequent MS/MS spectra yield quantitative reporter ions at m/z 114–117 along with amide backbone fragments. (Figure modified from Ross et al. 2004.)

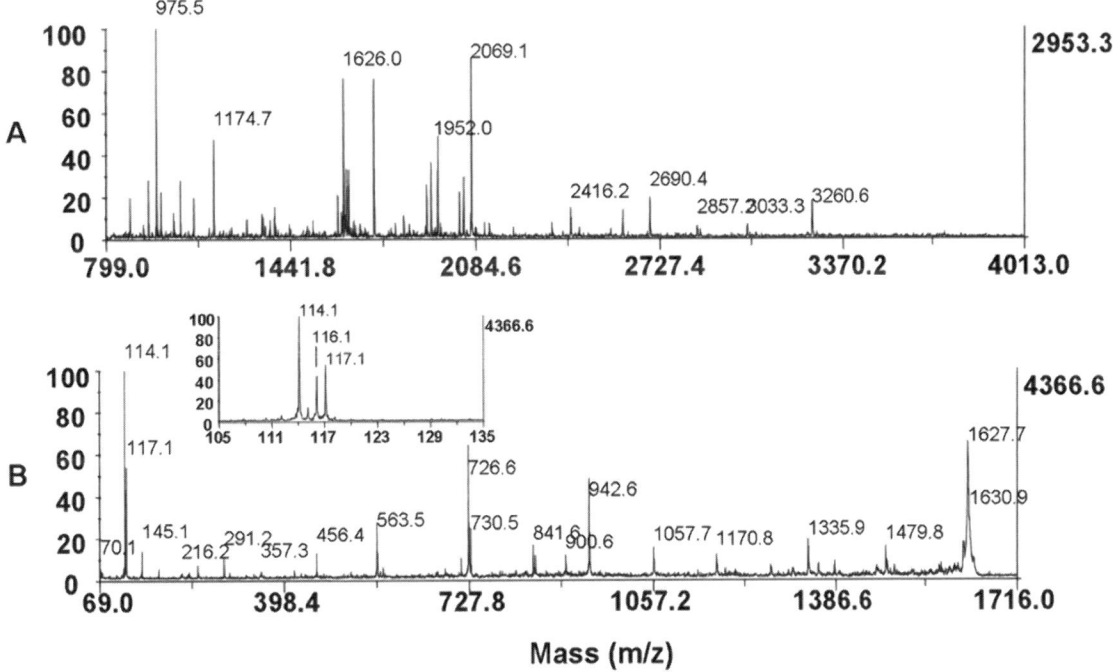

FIGURE 3. (A) A representative MS spectrum of iTRAQ-labeled peptides acquired on a MALDI TOF/TOF mass spectrometer. (B) A subsequent MS/MS spectrum of precursor ion m/z 1626.0 displaying reporter ion (inset) and backbone fragments.

of the MS/MS spectrum that contains the peaks corresponding to the iTRAQ reporter ions. The intensities of the reporter ions reflect the relative abundance of the peptide in each of the samples queried.

EXPERIMENTAL DESCRIPTION

In this experiment, iTRAQ will be used to compare the proteome of *Saccharomyces cerevisiae* under two different metabolic states: fermentation versus respiration. The goal is to determine the proteins involved in each of the activated pathways by measuring their relative abundances using the iTRAQ protocol. For the CSHL Proteomics course, yeast cells have been grown and lysed, and the soluble proteins have been digested with trypsin ahead of time. A summary of these steps is provided in Protocol 1. Figure 4 provides a diagram of the workflow that will be followed in the experiment. Briefly:

1. The peptides are labeled with different iTRAQ tags. With four iTRAQ tags (114, 115, 116, and 117) and only two samples, a double duplex experiment can be performed. Two aliquots of the peptides generated from the cells grown under fermentative conditions are labeled with tags 114 and 115. The peptides generated from the cells grown under respiration are labeled with tags 116 and 117.

2. After quenching the labeling reactions, the peptides are combined into one tube and cleaned up on a reversed-phase C18 cartridge to remove salts and by-products of the iTRAQ reaction.

3. The peptides are fractionated by isoelectric focusing (IEF).

4. The peptides are separated by reversed-phase chromatography and fractions are incrementally mixed with matrix and spotted by an automated MALDI plate-spotting robot.

5. The MALDI plates are analyzed with a 4700 MALDI TOF/TOF mass spectrometer.

6. The data is then analyzed using ProteinPilot software.

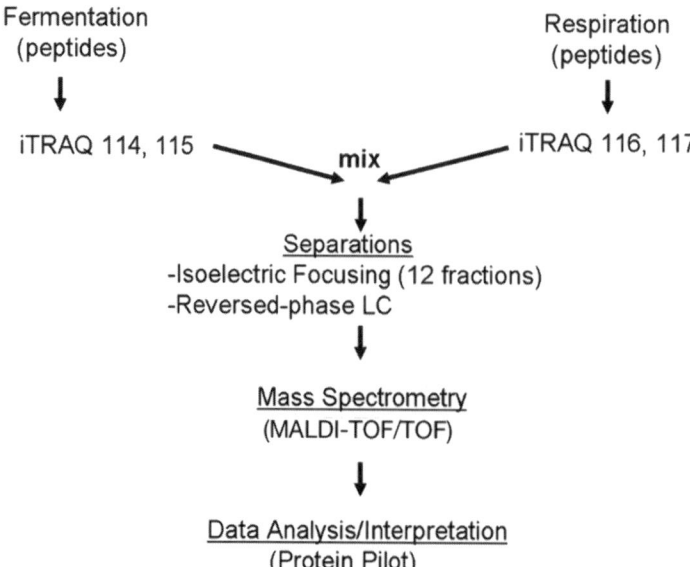

FIGURE 4. Workflow diagram for the yeast proteome experiment using the iTRAQ protocol.

PROTOCOL 1

Preparation of Peptides from Yeast Cells

MATERIALS

CAUTION: See Appendix 11 for appropriate handling of materials marked with <!>.

Reagents

Acetone, cold (chill at –20°C) <!>
Bradford protein assay reagents
200 mM Dithiothreitol (DTT) <!>
DNase/RNase solution (used in lysis solution)
    1 mg/mL DNase I
    0.25 mg/mL RNase
    500 mM Tris-Cl (pH 7.0)
    50 mM $MgCl_2$ <!>
200 mM Iodoacetamide <!>
Lysis solution
    12 mL CelLytic (Sigma-Aldrich C4482)
    150 µL DNase I/RNase solution
    3 Protease tablets (non-EDTA) (Roche 10946900)
    5 mM Tris-carboxyethyl phosphine (TCEP) <!>
500 mM Triethylammonium bicarbonate (TEAB)
Trypsin (sequence grade; Promega) <!>
7 M Urea/500 mM TEAB
Yeast cells (*S. cerevisiae* wild-type strain BJ1991)
YPD medium
    1% (w/v) yeast extract
    2% (w/v) peptone
    2% (w/v) *D*-glucose
YPG medium
    1% (w/v) yeast extract
    2% (w/v) peptone
    3% (w/v) glycerol

Equipment

Centrifuge
H_2O bath or heat block set at 37°C
Shaker
Vacuum evaporator (e.g., SpeedVac, Savant)

PROCEDURE

Growing Yeast Cells

1. For fermentative growth, cultivate *S. cerevisiae* wild-type strain BJ1991 on rich YPD. Inoculate a cell colony in 50 mL of liquid YPD and grow to an optical density (OD) of 1

($\sim10^7$ cells). Add 0.1 to 5 mL (volume depends on calculated doubling time and number of cells desired) of the starter culture to 1 liter of liquid YPD medium and grow to an OD of 1. Divide into 50-mL aliquots and pellet the cells by centrifugation (5000 rpm). This yields approximately 0.8–1 g of cells per 50 mL.

2. For growth under respiration conditions, grow a wild-type starter culture in rich YPD to an OD of 1–2 (as described in Step 1). Add 0.1 to 5 mL of the starter culture to 1 liter of liquid YPD medium and grow to an OD of 1 as described in Step 1. Pellet the cells (50-mL aliquots) by centrifugation (5000 rpm). Transfer the cells to YPG containing 3% (v/v) glycerol. Collect the cells by centrifugation from 50-mL aliquots (5000 rpm) after 16 hours. This yields approximately 0.8–1 g of cells per 50 mL.

The yeast cell pellets from Steps 1 and 2 can be stored frozen at –80°C.

Lysis of Yeast Cell Pellets

3. After allowing yeast cells to thaw, add 4 mL of lysis solution to 0.9 g of yeast cell pellet. Shake the mixture *gently* for 30 minutes at room temperature.

4. Remove cell debris by centrifugation at 4°C.

5. Collect the soluble protein. Using the Bradford method, determine the amount of protein recovered, and store the sample at –80°C until needed.

Precipitation of Protein

6. Add cold acetone to the protein sample at a 5:1 ratio (acetone/sample, v/v). Incubate it for 2 hours at –20°C.

7. Centrifuge the sample at 12,500g for 5 minutes at 4°C.

8. Discard the supernatant. Wash the pellet with cold acetone and centrifuge again.

9. Dry the protein pellet in a vacuum evaporator for 10 minutes to remove residual acetone.

10. Dissolve the pellet in 60 µL of 7 M urea/500 mM triethylammonium bicarbonate (TEAB).

    TEAB is used in place of ammonium bicarbonate because it has no primary amines, which would interfere with iTRAQ labeling.

Reduction, Alkylation, and Digestion of Proteins

11. Add 5 µL of 200 mM DTT to reduce the disulfide bonds. Incubate for 1 hour at room temperature.

12. Add 20 µL of 200 mM iodoacetamide (carbamidomethylation). Incubate the reactions for 1 hour in the dark at room temperature.

13. Quench the alkylation reaction by adding 20 µL of 200 mM DTT and incubating for 1 hour at room temperature.

14. Dilute the protein samples with 500 mM TEAB, so that the urea concentration is less than 2 M.

15. Add trypsin to a final 1:25 ratio (trypsin/sample). Incubate the mixture overnight at 37°C.

16. Store the peptides at –20°C.

NOTE: The iTRAQ reagent has some inherent instability and reactivity issues that should be understood before carrying out an iTRAQ labeling experiment. First, since the reagent is intended to modify peptides at primary amines, it is important to eliminate interfering background amines. Second, the reaction proceeds optimally between pH 8 and 9. To satisfy the first two conditions, triethylammonium bicarbonate (TEAB) is used to buffer the solution at about pH 8.5. This buffer replaces the more traditionally used ammonium bicarbonate, which reacts with iTRAQ. Third, the reagent is highly susceptible to hydrolysis. To combat this, the reaction is carried out in 70% ethanol, which slows the rate of hydrolysis enough to allow the primary amine reaction to go to completion. It pays to keep these considerations in mind, especially during preparation of the peptide samples. Avoid primary amines in sample preparation buffers, and keep the protein concentration of the sample as high as possible so that the amount of H_2O in the reaction is minimized during iTRAQ derivatization.

PROTOCOL 2

Labeling Peptides with the iTRAQ Reagent

The peptides will be labeled in what is known as a double duplex experiment. This means that two tags will be used for labeling peptides generated from fermenting yeast and two tags for labeling peptides generated from yeast induced to respire (Table 1; also see Fig. 4). This approach provides two measurements of each sample.

MATERIALS

CAUTION: See Appendix 11 for appropriate handling of materials marked with <!>.

Reagents

iTRAQ reagent kit (Applied Biosystems 4352135)
(Optional) 0.1% Trifluoroacetic acid (TFA) <!>
Tryptic peptide digests of *S. cerevisiae* (from Protocol 1, Steps 15 and 16)

PROCEDURE

1. Remove two aliquots of each tryptic peptide digest from the freezer. In the Proteomics course, each vial contains 30 µL of solution containing ~40 µg of total peptide.

TABLE 1. iTRAQ labeling strategy for protein quantification

iTRAQ Tag	Yeast Conditions
114	Fermentation
115	Fermentation
116	Respiration
117	Respiration

2. Remove one vial of each of the four iTRAQ tags. Allow approximately 5 minutes for the reagents to reach room temperature. Centrifuge each vial briefly to ensure the reagent is at the bottom of the tube.

3. Add 70 µL of ethanol (included with the kit) to each of the four iTRAQ reagent vials. Vortex each tube briefly to mix and centrifuge the tubes to ensure the mixture is at the bottom of the tube.

4. Add the contents of each of the reagent vials (now diluted with ethanol) to the appropriate sample vial. Vortex, centrifuge, and incubate the mixtures for 1 hour at room temperature.

   According to the manufacturer, the recommended reaction time is 1 hour. However, the incubation time can be reduced to 30 minutes, as the reaction will most likely be complete at that time.

5. (Optional) After 1 hour, most of the remaining reagent will be hydrolyzed. However, to quench the reaction and remove any remaining active reagent, add 100–200 µL of 0.1% TFA.

PROTOCOL 3

Peptide Clean-up Using a Reversed-phase Column

Before using isoelectric focusing (IEF) to fractionate the iTRAQ-labeled peptides, the sample needs to be cleaned up. The sample presently contains a high concentration of buffer (500 mM triethylammonium bicarbonate), excess reagents from the reduction and alkylation of the proteins, and degradation products of the iTRAQ reagents. All of these are incompatible with IEF. To remove them, the sample is passed over a reversed-phase C18 column.

MATERIALS

CAUTION: See Appendix 11 for appropriate handling of materials marked with <!>.

Reagents

Acetonitrile <!>
70% Acetonitrile/0.1% trifluoroacetic acid (TFA) <!>
iTRAQ-labeled peptide mixtures (from Protocol 2)
Reversed-phase C18 column (Phenomenex)
0.1% TFA

Equipment

Vacuum evaporator (e.g., SpeedVac, Savant)

PROCEDURE

1. To remove ethanol from the iTRAQ-labeled peptide samples, dry them down in a vacuum evaporator for approximately 20 minutes.

2. In the meantime, wet the reversed-phase C18 column with 1 mL of acetonitrile. Allow the solvent to filter through the column by gravity.

3. Condition the column with 1 mL of 0.1% TFA.

    It will take ~20 minutes to complete Steps 2 and 3.

4. Add 300 µL of 0.1% TFA to each tube of dried peptides (from Step 1). Combine all four of the samples into one tube.

5. Load the sample onto the reversed-phase C18 column. Allow the sample to pass through the particle bed by gravity.

6. Wash the column with 1 mL of 0.1% TFA.

7. Elute peptides from the column with 600 µL of 70% acetonitrile/0.1% TFA.

8. Dry the sample in a vacuum evaporator to a final volume of 100–200 µL.

PROTOCOL 4

Isoelectric Focusing of Peptides

So far, samples have been labeled with iTRAQ tags, and contaminants have been removed using reversed-phase column chromatography. Next begins the fractionation stage of the experiment, employing IEF as the first dimensional separation of the peptides. In this protocol, IEF is performed using an OFFGEL Fractionator (model 3100, Agilent Technologies) on immobilized pH gradient (IPG) strips with pH range 4–7.

MATERIALS

CAUTION: See Appendix 11 for appropriate handling of materials marked with <!>.

Reagents

iTRAQ-labeled peptide mixture (from Protocol 3)
Mineral oil (Bio-Rad)
Rehydration solution
> Prepare rehydration solution by diluting 25 µL of the stock solution of pH 4–7 ampholytes (GE Healthcare) to 10 mL with H_2O.

Equipment

Electrode wicks (Bio-Rad)
IEF fractionating device (Model 3100 OFFGEL Fractionator; Agilent Technologies)
IPG strips (pH 4–7, 13 cm) (GE Healthcare)
Tweezers

PROCEDURE

1. Dilute the iTRAQ-labeled peptide mixture with 1.9 mL of rehydration solution.

    Each well on the IPG strip holds 0.15 mL x 12 wells = 1.8 mL.

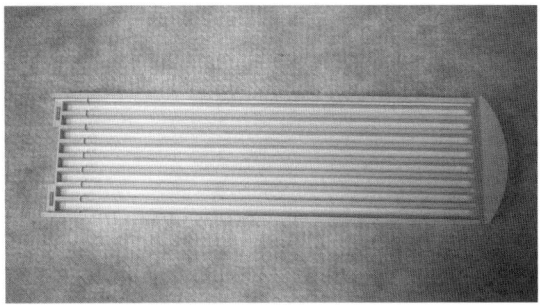

FIGURE 5. Proper orientation of the tray.

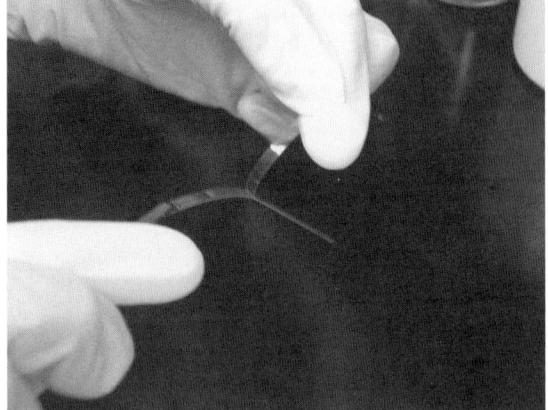

FIGURE 6. The end of the gel backing is held with forceps while the protective cover is peeled off the gel.

2. Place the gel tray (white with 8 lanes) on the bench so that the handle (curved end) is on the right (Fig. 5).

3. While wearing gloves, remove the protective backing from the IPG strip. Once the backing is removed, avoid touching the gel (Fig. 6).

4. Place the gel in one of the tray lanes with the *gel side up* and the anode end of the strip (marked "+") on the left. When using gels from GE Healthcare, the writing on the gel will appear upside down and backward (Fig. 7). Make sure that the gel abuts the left side of the tray.

5. Keeping the right side of the frame (long piece with 12 wells) slightly elevated, place the left side of the frame against the mechanical stop (Fig. 8A). Lower the frame and press until it snaps (Fig. 8B). Be careful not to move or crimp the gel strip during this maneuver.

6. Pipette 20 µL of rehydration solution into each well. Do not touch the gel when doing so. Gently tap the tray on the bench top so that the rehydration solution reaches the gel. It is not necessary that the entire gel surface is covered.

7. Using tweezers, wet one electrode pad, or wick, with rehydration solution.

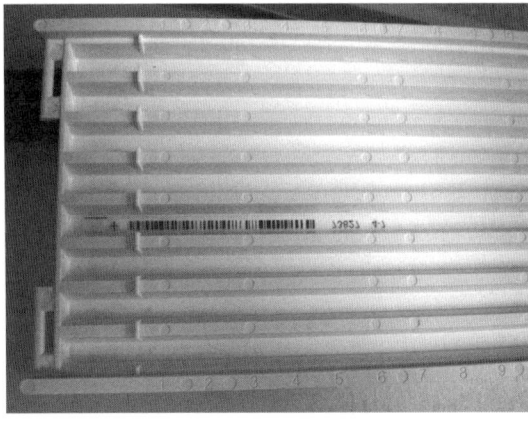

FIGURE 7. The gel is placed in the lane with the gel facing up.

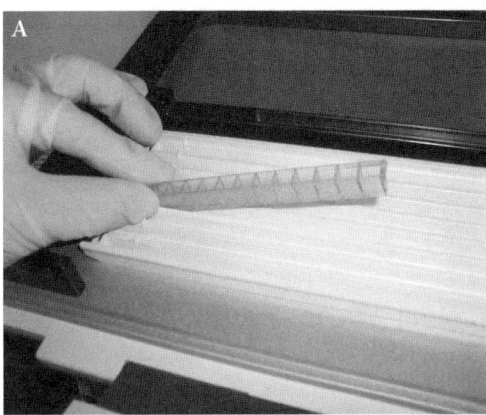
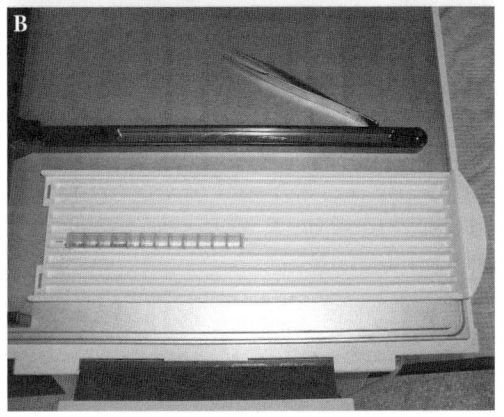

FIGURE 8. (A) The right side of the frame is elevated while the left is placed against the mechanical stop. (B) The frame is snapped in place above the gel with the left side abutting the mechanical stop.

8. Place the wetted pad, or wick, on one protruding end of the strip, making sure that there is no gap between the pad and the frame (Fig. 9). Note: It may be necessary to cut the pad along its length to fit into the lane. This should be done ahead of time.
9. Place a wetted pad on the opposite end of the strip in the same manner.
10. Place another wetted pad on top of each of the two previously placed pads on each end. Each end should now have two wetted pads stacked on top of each other.
11. Allow the gel to swell for approximately 15 minutes.
12. Add 150 μL of sample (prepared in Step 1) to each of the 12 wells.

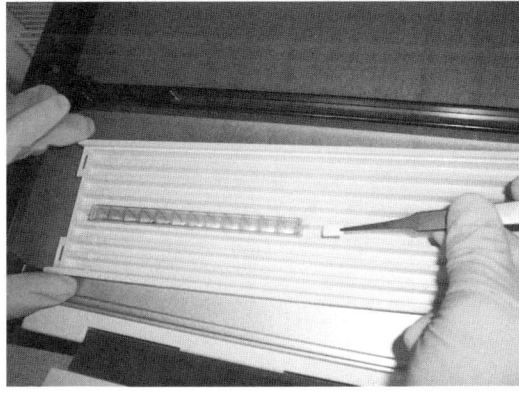

FIGURE 9. Placement of paper wicks on top of the edge of the IPG strip and abutting the frame.

FIGURE 10. Placement of the cover slip above the frame.

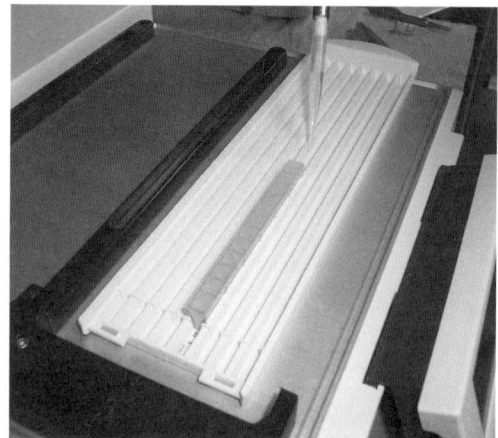

FIGURE 11. Mineral oil is dispensed in the lane on both sides of the frame.

13. Place the cover seal over the frame and press down gently on each well to secure the seal to the frame (Fig. 10). Be careful not to move the frame in the tray.

14. Apply another 10 µL of rehydration solution onto the electrode pads at each end of the strip. Be careful not to move the pads.

15. Place the tray onto the instrument platform with the anode end ("+") on the left. The curved end of the tray should be on the right.

16. Pipette 200 µL of mineral oil onto the anode end ("+") of the strip. Pipette 1 mL of mineral oil into the lane on the cathode side of the strip (curved end of the tray) (Fig. 11).

17. After one minute, reapply 200 µL of mineral oil to both ends of the strip. (The oil should not extend higher than half the height of the tray grooves.)

18. Install the fixed electrode by placing the two tabs on the electrode into the slots on the left side (anode, +) of the tray (Fig. 12A). Rotate the fixed electrode down into position over the electrode pads (Fig. 12B). Push down until the electrode clicks into place. Without lifting the tray, slide it into the anode connector (Fig. 12C).

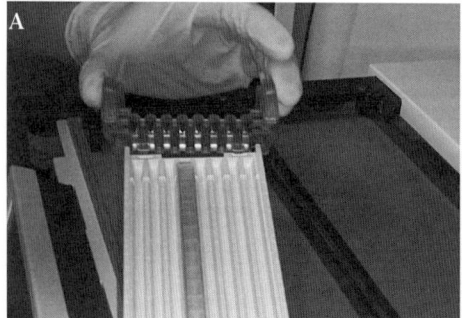  

FIGURE 12. The tabs on the fixed electrode are inserted into the slots on the left side of the tray (A) and the electrode is rotated downward (B). (C) The tray is then slid into the anodic connector.

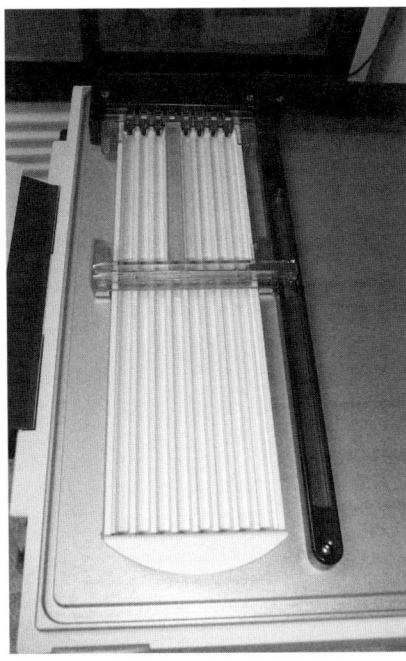

FIGURE 13. The movable electrode is placed abutting the right side of the frame.

19. Insert the movable electrode onto the cathode end of the tray such that the electrode makes contact with the electrode pads and abuts the cathode end of the frame (Fig. 13).

20. Close the lid on top of the off-gel apparatus.

21. Load the program. The off-gel device is preprogrammed with methods for various applications. For example, the device contains a method that is optimal for running peptides on 13-cm IPG strips. The method is loaded using the handheld controller on the front of the device.

    a. Select the tray position: Tray I or II.

    b. From the menu options on the screen, select "Method," then "Load," and select the method "OG12PE00." You will see a table of parameters including the targeted volt hours and maximal voltage, current, power, and time thresholds. It also lists the holding potential and maximum current used to keep the peptides focused once the desired voltage hours have been achieved.

    c. Select "Done."

    d. Select "Start."

       The fixed electrode will light up once the method has begun. When finished, it will blink on and off. The method will typically run for 4 to 12 hours.

22. To extract the peptides from the 12 wells, remove the solution above the gel with a pipette and dispense into a properly labeled microcentrifuge tube.

23. Inject each of the 12 IEF fractions onto a reversed-phase HPLC column to separate the peptides. The peptides are spotted and mixed with matrix automatically by a robot as they elute off of the LC column. Acquisition of the data on the 4700 MALDI TOF/TOF is covered in Experiment 3, and data analysis of the data with Protein Pilot is covered in Experiment 8.

REFERENCES

Choe L., D'Ascenzo M., Relkin N.R., Pappin D., Ross P., Williamson B., Guertin S., Pribil P., and Lee K.H. 2007. 8-Plex quantitation of changes in cerebrospinal fluid protein expression in subjects undergoing intravenous immunoglobulin treatment for Alzheimer's disease. *Proteomics* **7**: 3651–3660.

Gygi S.P., Rist B., Gerber S.A., Turecek F., Gelb M.H., and Aebersold R. 1999. Quantitative analysis of complex protein mixtures using isotope-coded affinity tags. *Nat. Biotechnol.* **17**: 994–999.

Ong S.E. and Mann M. 2005. Mass spectrometry-based proteomics turns quantitative. *Nat. Chem. Biol.* **1**: 252–262.

Ong S.E., Blagey B., Kratchmarova I., Kristensen D.B., Steen H., Pandey A., and Mann M. 2002. Stable isotope labeling by amino acids in cell culture, SILAC, as a simple and accurate approach to expression proteomics. *Mol. Cell. Proteomics* **1**: 376–386.

Ross P.L., Huang Y.N., Marchese J.N., Williamson B., Parker K., Hattan S., Khainovski N., Pillai S., Dey S., Daniels S., et al. 2004. Multiplexed protein quantitation in *Saccharomyces cerevisiae* using amine-reactive isobaric tagging reagents. *Mol. Cell. Proteomics* **3**: 1154–1169.

EXPERIMENT 8

Analysis and Validation of Tandem Mass Spectra*

The combination of mass spectrometry and genome-assisted data analysis has revolutionized proteomics. Large numbers of proteins and modified amino acids can now be rapidly identified from small amounts of material. Computational analysis and interpretation of the acquired mass spectrometry data play an essential role in these proteomics experiments. In Experiments 1 through 7 of this manual, precursor spectra (MS) were used to measure the masses (m/z) of peptides from trypsin-digested proteins. Selected peptides were fragmented by collision-induced dissociation (CID) to generate fragmentation spectra (MS/MS). During these experiments, the intensity and m/z values of the precursor and fragmentation ions were saved into a data file. In this experiment, the data in these files are extracted into a format that can be read by database search programs. The programs compare the experimental data collected from the mass spectrometry experiment to the theoretical masses of peptides and fragmentation ions from proteins in a database. Using various mathematical and statistical scoring approaches, the database search programs identify peptide sequences in the database that best match the experimental data. Finally, the programs generate a list of proteins predicted to be in the analyzed sample. Evaluating the search results and validating the identified proteins and peptides is a major challenge. Data analysis requires understanding the strengths and weaknesses of the search programs along with a basic understanding of how to interpret and validate peptide fragmentation spectra.

In tandem-mass-spectrometric sequencing of a peptide, information about the peptide sequence is contained in the product ion or MS/MS spectrum. Low-energy fragmentation of ionized tryptic peptides occurs primarily at the amide bonds along the peptide backbone, generating a series of fragmentation or product ions (Fig. 1). Several models have been proposed that describe the fragmentation chemistry, including the mobile proton model and the pathways in competition model (Dongré et al. 1996; Wysocki et al. 2000; Paizs and Suhai 2005). The sequence of the peptide is determined by comparing the differences among the masses of the product ions with the known masses of the individual amino acid residues (see Appendices 5 and 6).

A standardized nomenclature has been adopted to describe the different product ions observed in fragmentation spectra (Fig. 2). The a-, b-, and c-ions all contain the amino terminus of the peptide, while the x-, y-, and z-ions all contain the carboxyl terminus. Under low-energy CID, the major amino-terminus ion series is the b-ion series and the major carboxy-terminus ion series is the y-ion series. The a-, c-, x-, and z-ion products are typically not produced under low-energy CID and are only observed under high-energy CID and other fragmentation methods, such as electron transfer dissociation (ETD). For both the b- and y-ion series, the m/z difference between adjacent ions is equal to the mass of the amino acid residue at that position (Fig. 1). For doubly charged peptides, the b- and y-ion series are complementary as the cleavage of the protonated amide bonds can generate

*Several sections contributed by Rebecca A. Bish (*Department of Cancer Biology and Genetics, Memorial Sloan-Kettering Cancer Center, New York, New York 10021*) and Eric S. Simon (*Department of Biological Chemistry, University of Michigan, Ann Arbor, Michigan 48109*).

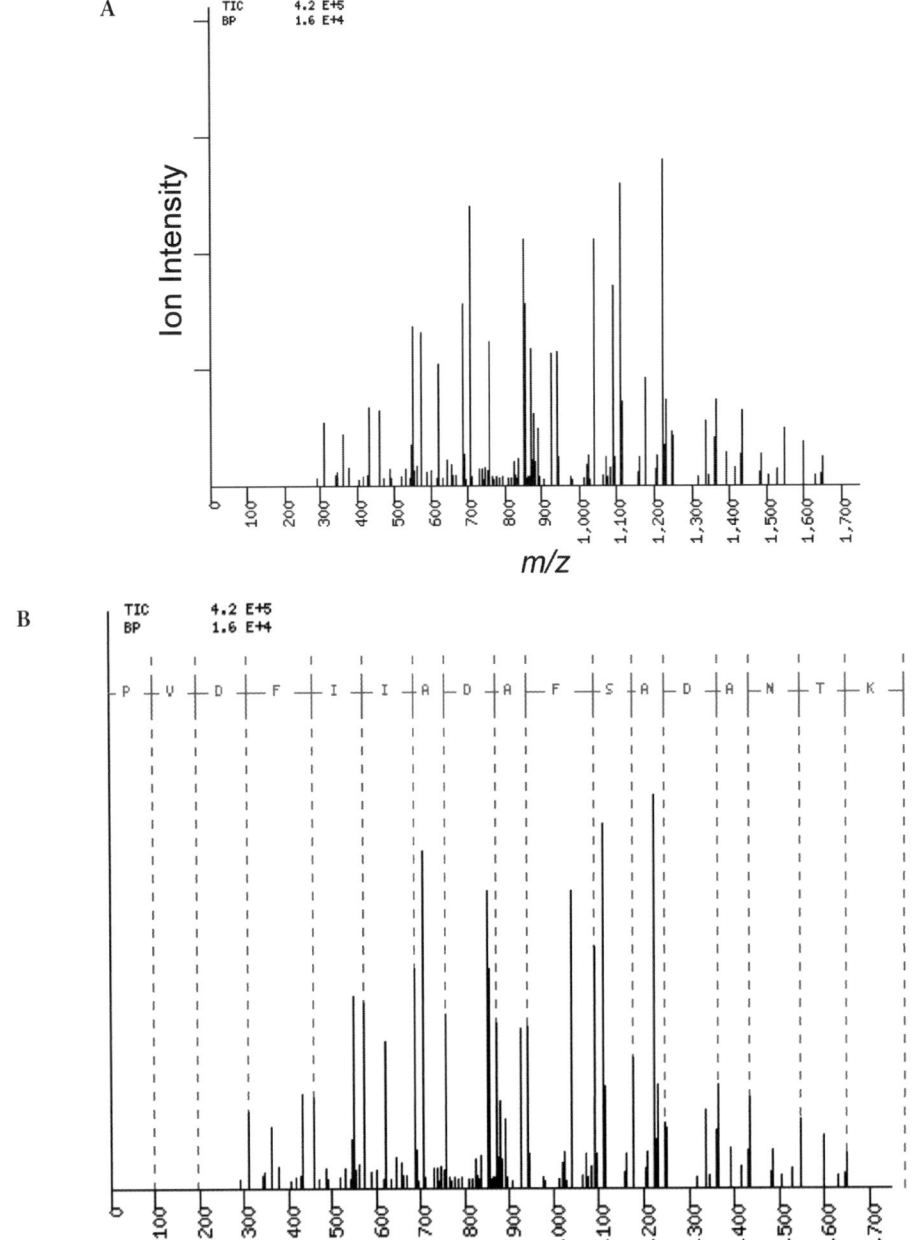

FIGURE 1. (A) Unannotated MS/MS or product spectrum. A fragmentation spectrum acquired for a doubly charged precursor ion (m/z 898.9) from a trypsinized yeast cell lysate. The x-axis shows the mass of the ions as the mass to charge ratio (m/z), and the y-axis is the abundance of the ion measured in ion intensity. The spectrum contains peaks from the peptide's fragmentation ions along with background noise. Typical of good quality data, the spectrum shows a broad range of ions that are significantly higher in intensity than the noise peaks. In addition, the intense ions appear to form a symmetrical pattern. (B) Theoretical b-ions from a peptide hit. The precursor and fragmentation ion data were submitted to a Sequest database search against the yeast proteome. The top scoring peptide was PVDFIIADAFSADANTK from the Pgk1 protein (Cn = 6.9). The program predicted the precursor was a doubly charged ion (z = $^+2$). The theoretical b-ions from the peptide are superimposed on the MS/MS spectrum showing which fragment ions might be b-ions. The mass difference between adjacent b-ions corresponds to the residue mass of an amino acid (see Appendices 5 and 6). (C) Theoretical y-ions from a peptide hit. The theoretical y-ions from the PVDFIIADAFSADANTK peptide are superimposed on the MS/MS spectrum showing which fragment ions might be y-ions. The mass difference between adjacent y-ions corresponds to the residue mass of an amino acid (see Appendices 5 and 6). (D) Theoretical b-y ions from a peptide hit. The MS/MS spectrum's fragmentation ions are labeled with the theoretical b-y ions from the peptide PVDFIIADAFSADANTK. For a good database match to a multiply charged precursor ion, most of the intense ions should match the predicted fragmentation ions. (E) Summary of peptide's match to the MS/MS spectrum. This method of interpretation is commonly used to represent which b-y ions from a peptide are detected in the MS/MS spectrum. The calculated m/z values of the b-ions are shown above the amino acid sequence and the calculated m/z values of the y-ions are shown below the sequence. All the values are singly charged and are the monoisotopic masses. The figure also highlights the complementarity of the b- and y-ion series. As a b- or y-ion is identified, its complementary ion can be calculated using the precursor ion mass. For evaluating a peptide hit to an MS/MS spectrum, detecting complementing ion pairs is one indication of an accurate database hit.

ANALYSIS AND VALIDATION OF TANDEM MASS SPECTRA 119

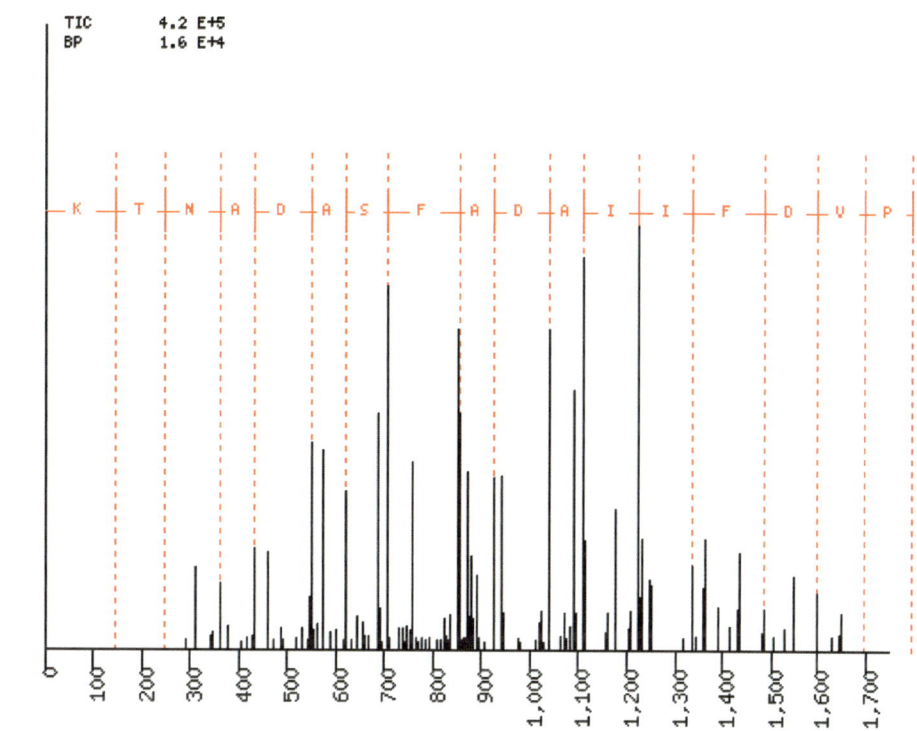

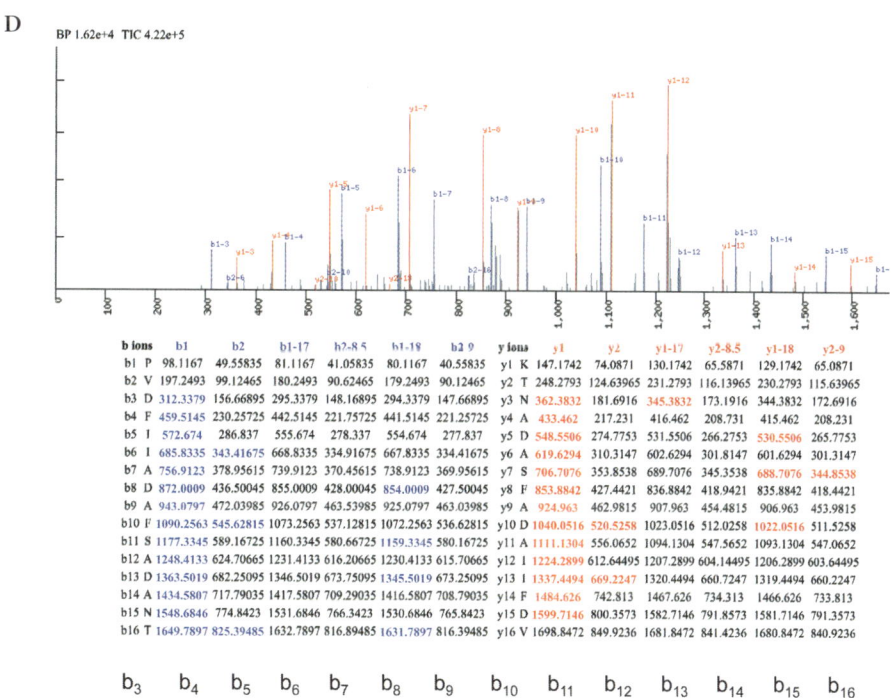

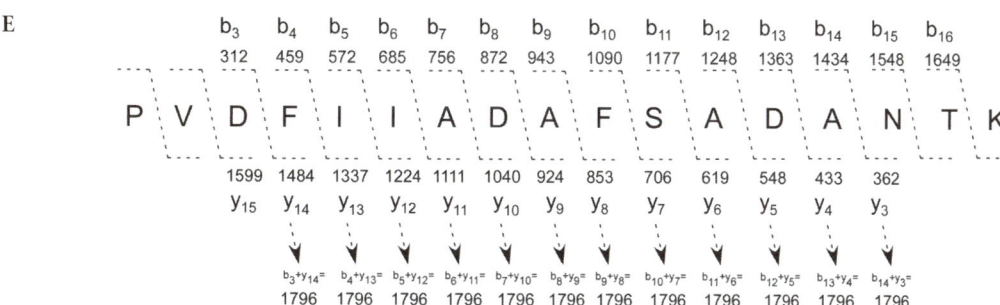

FIGURE 1. (See facing page for legend.)

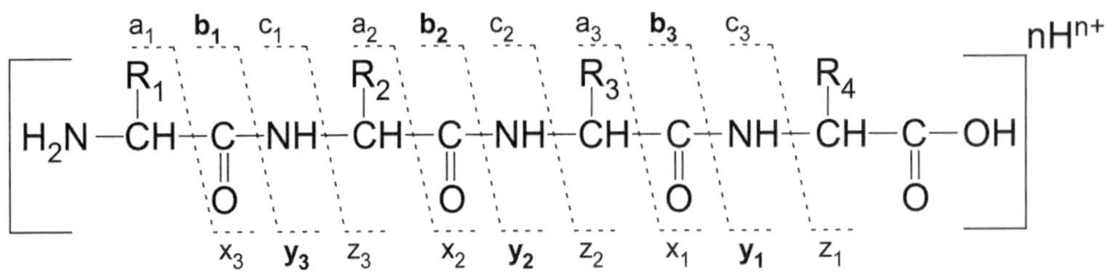

FIGURE 2. Nomenclature for fragmentation ions generated from a peptide. The diagram shows two classes of fragmentation or product ions that are created from a protonated peptide. One class contains the amino terminus (a_i, b_i, c_i) while the other class contains the carboxyl terminus (x_i, y_i, z_i). For both classes, the fragmentation can occur at three different positions along the peptide backbone. The b- and y-ions (bold) are the dominant fragmentation ions created by low-energy CID. For the b-y ions, the fragmentation occurs at the peptide's amide bonds. The diagram illustrates the complementarity of fragment ions. For example, the sum of $b_1 + y_3$ ion masses equals the precursor ion mass.

both a b-ion and y-ion (Figs. 1 and 2). The b- and y-ions are derived from single fragmentation reactions that occur from a population of precursor ions protonated at the different amide bonds. Recognizing the members of the b- and y-ion series in the product spectrum and the calculation of the residue masses is the fundamental process of de novo amino acid sequence determination.

The de novo interpretation of tandem spectra is not straightforward. The b- and y-ions do not stand out upon casual observation (Fig. 1A). The relative abundances of product ions typically vary extensively, with some product ions dominating the spectrum while other product ions are undetectable (Fig. 1B,C). The variation reflects the differences in the amide bonds due to the properties of the amino residues flanking the bonds and the position of the bonds in the peptide sequence (see Protocol 3). Losses of neutral molecules from the precursor and b- and y-ions reduce the intensities of the fragmentation ions. For tryptic peptides, the loss of water (–18 Da) from Ser, Thr, Glu, and Asp residues and ammonia (–17 Da) from Asn, Gln, Arg, and Lys residues are frequently observed. These neutral losses generate additional product ions from the loss of –17 or –18 Da from the b- and y-ions (Fig. 1D). Like most analytical techniques, mass spectrometry data contain background noise from the sample and the mass spectrometry system, which adds to the overall complexity and uncertainty. In a limited number of examples, the combined residue mass of two amino acids may equal the residue mass of a single amino acid. For example, the combined residue masses of Asp-Ala, Val-Ser, and Gly-Glu all equal the residue mass of Trp (186 Da). Finally, the residue masses of isoleucine and leucine are identical (113 Da) and cannot be distinguished using low-energy CID. Given these complications, the de novo interpretation of tandem spectra can be complicated and time consuming. The lack of information in the MS/MS spectrum during de novo analysis frequently leads to an incomplete peptide sequence.

For proteomic experiments, the de novo interpretation of tandem mass spectra is typically bypassed and the initial interpretation of spectra is done using database search programs (Fig. 3). The programs compare the experimental mass spectrometry data to theoretical data derived from protein sequences in a database. Database search algorithms rapidly process large numbers of MS/MS spectra and return a list of peptide sequences that match the MS/MS spectra. From the peptide sequences, the programs build a list of proteins predicted to be in the sample. Evaluating the accuracy of these lists is one of the most difficult challenges in proteomics

The database search programs first identify strings of amino acids from proteins in the database whose mass is equivalent to the precursor ion's mass. Then, the theoretical masses of predicted product ions of each of these peptides are compared to the actual ion masses and intensities in the MS/MS spectrum. The mathematical and statistical methods used for comparing and scoring the theoretical values and experimental data constitute the primary differences among the different database search programs (Sadygov et al. 2004). Once a candidate peptide sequence has been iden-

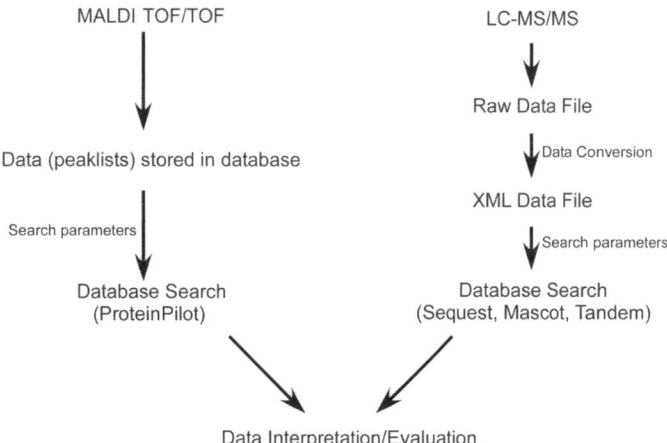

FIGURE 3. Flow diagram showing the process for analyzing MALDI TOF/TOF and LC-MS/MS data.

tified, the b- and y-ion series in the experimental spectra can be identified (Fig. 1B–E). The database search algorithms are designed to provide a score or statistical value to quantify how significantly each peptide sequence in the database matches the experimental data. The programs rank the peptide scores to show the best peptide match. Ideally, each protein is identified by multiple, independent peptide matches. However, in reality, a large percentage (>50%) of the identified proteins are typically based on a single peptide match to an MS/MS spectrum. For these "one hit wonders," the quality of the spectral data or the match of the peptide sequence to the MS/MS spectrum may be marginal. Confirmation of the match requires careful manual inspection of the MS/MS spectrum and determination that the predicted product ions support the fragmentation data.

PROTOCOL 1

Analyzing LC-MS/MS Data Using the Global Proteome Machine

MATERIALS

Binary data file generated from an LC-MS/MS experiment
Computer running Windows XP or Linux

PROCEDURE

Converting Native Mass Spectrometry Data Files to an XML Format

Most mass spectrometer vendors encode the data files in a propriety binary format. The m/z and intensity values from the precursor (MS) and product (MS/MS) spectra need to be extracted from these native files into a text format that can be read by the database search programs. XML is a simple, flexible text format derived from the standardized general markup language (SGML), originally designed for electronic publishing. Two groups have independently proposed a common file format based on XML (mzXML and mzData) for representing native mass spectrometry data. This common format allows data from different instrument manufacturers to be analyzed by a database search algorithm. Recently, the two XML formats were reconciled and a new format was created, mzML, which is intended to replace the mzXML and mzData formats (Deutsch 2008). Most native data files can be converted to an XML format using software applications provided by the instrument manufacturers.

In the course, a ThermoFisher LTQ mass spectrometer is used for LC-MS/MS experiments. The acquired data are recorded in a native, binary file with the "RAW" extension. To analyze the data using the database search programs, the RAW file needs to be converted to either XML or some other text file format. We use the publicly available ReadW.exe program to convert the RAW files to the XML file format mzXML. Software is also publicly available for converting between XML and ASCII formats (http://www.proteomecommons.org/).

1. Install the ReadW.exe program in the "C:/WINDOWS/system32 folder" on a PC running ThermoFisher's Xcalibur software.

    The ReadW.exe program can be downloaded from either the open source software website SourceForge (http://sourceforge.net/) or the Institute for Systems Biology (http://tools.proteomecenter.org/ReAdW.php). Because the ReadW program depends on Windows-only vendor libraries from ThermoFisher, this code will only work under Windows with ThermoFisher's XCalibur software installed. Other conversion programs are used to convert the native files from other mass spectrometer vendors. The mxSTAR.exe programs convert SCIEX/ABI's Analyst files. The mzBruker.exe program converts native Bruker format, and MassWolf.exe converts Micromass's MassLynx files to the mzXML format.

2. Open the command prompt on the PC by clicking on "Start" > "Run" and then typing "cmd."

3. Change directories to the folder containing your RAW file by typing the command prompt "<cd c:\xcalibur\data>."

4. Type the command "<readw.exe filename.raw>" to run the conversion program.

5. When the program has finished, an mzXML file with the same name as the original file will be found in the "C:\Xcalibur" directory.

Searching Tandem Mass Spectrometry Data

Numerous database search programs have been developed to compare the measured values of the precursor ion and its fragmentation ions to the theoretical masses of peptides and their fragmentation products derived from protein sequences in a database. Most mass spectrometer vendors offer a commercially licensed database search program for processing the data with their mass spectrometers. Three of the most widely used search algorithms are Sequest, Mascot, and X!Tandem (Eng et al. 1994; Perkins et al. 1999; Craig and Beavis 2004). All three programs can be used to analyze native data from any mass spectrometer once the data file has been converted to an XML or an ASCII format. SEQUEST, which was the first program to take unedited tandem mass spectra and compare the data to sequences in a protein database (Eng et al. 1994), is typically used by operators of ThermoFisher mass spectrometers. The SEQUEST algorithm uses a cross-correlation function to assess the similarity between the experimental and predicted spectra. Peptide sequences matching the spectrum are ranked by a cross-relation score (Cn). Mascot and X!Tandem use a probability-based scoring algorithm to derive from a protein database the most likely peptide sequence matching the experimental spectrum (Pappin et al. 1993; Perkins et al. 1999; Craig and Beavis 2003; Fenyo and Beavis 2003). Operators of SCIEX/ABI instruments typically use Mascot. X!Tandem or Tandem (http://www.thegpm.org/) was the first open source database search algorithm that was freely available and was being used by a growing number of investigators.

While the database search algorithms match acquired tandem mass spectra to peptide sequences in a database, the peptide sequences must be assembled into proteins that represent the initial sample. Early generation assembler applications such as SEQUEST Summary and Autoquest used a simple scoring threshold when assembling peptides into proteins (McCormack et al. 1997; Link et al. 1999). Second generation applications such as DTASelect and INTERACT applied layers of filters at the peptide and protein level when assembling peptide sequences into a list of proteins (Han et al. 2001; Tabb et al. 2002). However, this process was complicated by nonunique peptide sequences that match multiple proteins in a database. In addition, the wide range of scoring confidences for peptide matches increased the ambiguity in the final list of proteins. More recently, the probability-based method of ProteinProphet, and peptide-centric approaches such as Isoform Resolver and parsimony analysis have been developed to construct protein profiles from peptide lists generated by spectra search algorithms (Nesvizhskii et al. 2003; Resing et al. 2004; Yang et al. 2004; Zhang et al. 2007).

Installing the GPM on a PC Running Windows

The global proteome machine (GPM) is an open-source interface that uses the X!TANDEM database search algorithm to conduct database searches on data files from mass spectrometry experiments. It was the first open source database search program. Data for analysis by the GPM can be submitted over the internet, although this option is generally too slow for the large files generated by LC-MS/MS experiments. Since the release of X!Tandem, other open source database search programs have become available, including the open mass spectrometry search algorithm (OMSSA) sponsored by NCBI (Geer et al. 2004) and MyriMatch by David Tabb's laboratory at Vanderbilt University (Tabb et al. 2007).

At the CSHL Proteomics course, we use the Sequest, Mascot, and X!Tandem programs for analyzing LC-MS/MS data. Since the GPM and X!Tandem are open source and freely available, we install them on the students' laptop computers. The students are encouraged to compare the results from all three programs on the same sets of data. Most students in our course find it convenient to use the GPM installed as a local copy on a personal computer. Unfortunately, there is no version of the GPM for the Macintosh operating system.

1. Go to http://www.thegpm.org.
2. On the left-hand side of the page, under the heading "Download," click the link for the ftp site.
3. Click the directory "Projects."
4. Click the directory "GPM."
5. Click the directory "Current_release" or "GPM-xe-installer." Download the latest version of the software to your computer.

    Software for the GPM is updated regularly. Check for the most current software releases for your computer and operating system and then download the most recent version of GPM. The download typically includes several commonly used protein databases from model prokaryotic and eukaryotic organisms. If you work on an organism whose database is not included, you can download and install whatever databases you want.

6. Use WinZip to extract all files to a folder in your Program Files directory.
7. To run the GPM, enter the Program Files\GPM folder and click on the GPM Manager.exe icon.

    Drag this icon onto your desktop if you wish to create a shortcut.

Setting Up a Simple GPM Database Search of an LC-MS/MS Experiment

This section describes typical settings for an X!Tandem database search of LC-MS/MS data acquired using a ThermoFisher LTQ (Fig. 4). In most cases, the default values are a good place to start. We

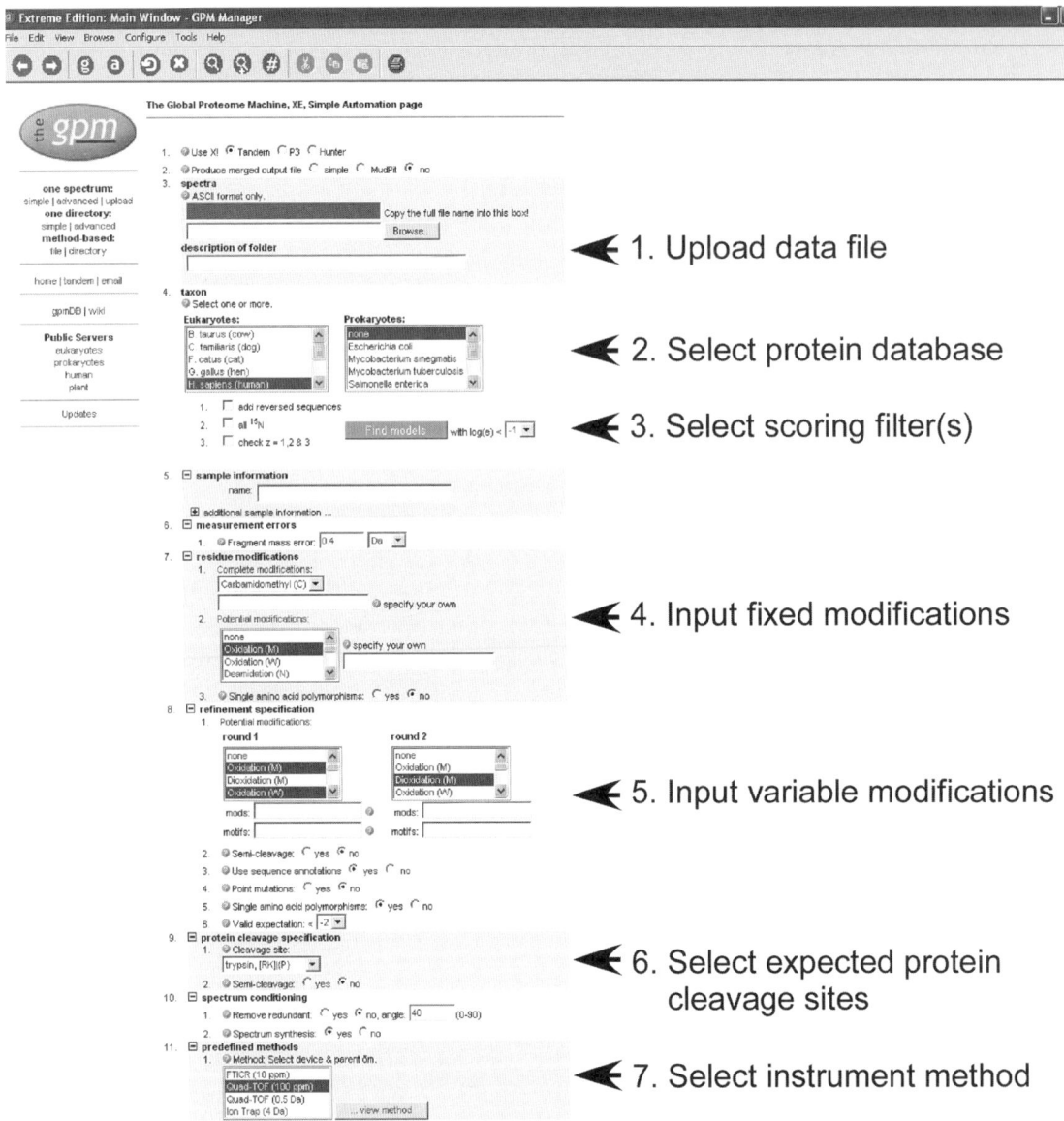

FIGURE 4. Parameter's page for setting up a Tandem search of acquired MS/MS data against a protein database. The typical order for inputting the required search information is shown on the right. All search engines (e.g., Sequest, Mascot, X!Tandem, etc.) require users to input similar information before running the database searches. Search engines will have different parameters for the scoring filter(s) based either on a probability value, false–positive rate, unique scoring values, or the proteolytic cleavage sites of the peptide.

have empirically determined that these recommended settings return good results for the LTQ instrument. However, these settings will vary significantly depending on which mass spectrometer you use. If you use the GPM to analyze your own data, you should adjust the settings as appropriate for the mass spectrometer used to acquire the data. If you use a proteomics or mass spectrometry facility, the manager of the facility should be able to help you determine the appropriate settings. It can also be helpful to run multiple searches on the same data set with different parameters for optimization purposes.

While this section describes the parameters for starting a database search using the GPM and the X!Tandem algorithm (Fig. 4), the input parameters will be identical for almost all other search algorithms (such as Sequest and Mascot). Users will be required to input an XML or ASCII version of their mass spectrometer's data file, select a protein database to search, select the protease used to digest the protein sample, select prefiltering parameters to remove redundant or poor quality spectra, input the expected mass accuracy of the acquired data, input any fixed or variable amino acid modifications, and input the scoring criteria for accepting a peptide sequence matching an MS/MS spectrum.

It is becoming common to compare the acquired mass spectrometry data to a true database as well as a decoy or false protein database. The false database is typically created from the true protein database by either reversing or randomly shuffling the amino acids for each protein. The actual and decoy databases are concatenated together and the mass spectrometry data file is simultaneously compared to the concatenated protein databases. From the false database, a false discovery rate (FDR) can be calculated for the experiment. The FDR estimates the percentage of identified proteins that are false hits. In describing the setup for a database search (below), we have chosen the GPM and X!Tandem since they are both publicly available and widely used by a number of groups.

1. Convert the ThermoFisher RAW file to an mzXML file using the ReadW.exe program as described in the first section of this protocol.

    This converts the native mass spectrometry data files to an XML format.

2. Under the heading "one spectrum" on the left-hand side of the page, click "advanced."

3. In the box labeled "spectra," browse to find your mzXML data file (Fig. 4).

4. Under the heading "taxon," browse to find the appropriate organism. In the CSHL course, most of the experiments are done using the yeast *S. cerevisiae* (Fig. 4).

    Excessively large protein databases may overwhelm a database search engine's ability to discriminate correct sequences from incorrect ones. Increasing the number of protein sequences searched against the MS/MS data will amplify the number of mathematical models to be tested. Unnecessarily large protein databases will increase the number of random or stochastic matches. As a consequence, correct identifications may be obscured by random matches. Select a protein database that is appropriate for the source of the proteins in the samples. Limit the search databases to a specific species or groups of species. A smaller database will also decrease the time to perform the database search. Add common protein contaminants (trypsin and human keratins) to the protein databases that are routinely found in samples.

5. Under the heading "measurement errors," change the following settings: Parent mass error should be +4 or –2 Da (Fig. 4).

    Be sure to change the units from ppm to Da. The measurement errors are dependent on the instrument being used for the tandem mass spectrometry experiment. These values are typically used for ion trap instruments. X!TANDEM allows the user to set asymmetric values for the parent ion mass error, to compensate for the asymmetric systematic mass errors generated by some types of mass spectrometers, particularly the ion trap instruments. TOF and FTICR mass analyzers will have smaller measurement errors.

6. Under the heading "signal processing," change the following settings: Minimum parent M+H should be 300 and the minimum peaks should be 10.

7. Add any desired posttranslational modifications (Fig. 4).

    Input any known or suspected modifications. Two types of modifications are entered. Fixed modifications are applied universally to every instance of the specified residue(s) or terminus. For most of the experiments in this manual, Cys residues have been reduced with DTT and alkylated with IAA, creating a carbamidomethyl on Cys residues, which means that all calculations will use 160 Da (157 Da + 57 Da) as the mass of Cys residues.

    Variable modifications are those which may or may not be present in the protein sample. The search algorithms will query all possible arrangements of the variable modifications. Common variable modifications include oxidized Met (+16 Da), deamidation of Asn or Gln (+1 Da), and phosphorylation of Ser, Thr, or Tyr (+80 Da) residues. There are numerous other potential modifications that can be searched (Creasy and Cottrell 2004).

For protein samples prepared using solutions containing urea, carbamylation (+43 Da) of the proteins' amino terminus and Lys, Arg residues are frequently observed, especially when the samples are heated. Isocyanatic acid, a urea break-down product, covalently reacts with amino groups. Database searches with a variable modification of +43 Da at the amino terminus and Lys, Arg residues will often detect carbamylated peptides.

As a caution, a single variable modification will generate many possible mathematical models to be tested. Multiple variable modifications cause the number of mathematical possibilities to increase geometrically. The database search will take considerably longer with more variable modifications. More importantly, the increased number of mathematical possibilities increases the number of random matches. Excessive protein variable modifications may overwhelm a database search engine's ability to discriminate correct sequences from incorrect ones. Vigilant data analysis is required to validate the accuracy of the modified peptide hit (see Protocol 3).

8. Click "Find models" to start the search.

Setting Up a MudPIT GPM Database Search

In the MudPIT experiment, a trypsin-digested, whole-cell lysate was fractionated using multidimensional HPLC and each fraction was analyzed by ESI-MS/MS (Experiment 6). Since each MudPIT fraction generates a data file, multiple mass spectrometry data files were created for the single sample. For X!Tandem to analyze and process the data, the mzXML-converted data files are placed in a single folder on the computer. X!Tandem sequentially searches the data files against a protein database and the results are merged into a single output file.

1. Convert all of the RAW files to mzXML files using the ReadW.exe program as described earlier.
2. Place all of the mzXML files from a single MudPIT run into a separate folder. This folder should not contain any mzXML files that are not to be included in the search.
3. In the GPM manager software, under the heading "one directory" on the left-hand side of the page, click "advanced."
4. Next to the heading "Produce merged output file," check the box for MudPIT.
5. Tell the GPM program which files to search in a combined MudPIT run.
6. Under the heading "taxon," browse to find the appropriate organism. In the CSHL course, most of the experiments are done using the yeast *S. cerevisiae*.
7. Under the heading "measurement errors," change the following settings: Parent mass error should be +4 or –2 Da.

    Be sure to change from ppm to Da.

8. Under the heading "signal processing," change the following settings: Minimum parent M+H should be 300 and the minimum peaks should be 10.
9. Add any desired posttranslational modifications.
10. Click "Find models" to start the search.

    The output from this search will be an individual results file for each fraction, plus a combined results file for the entire MudPIT run.

Preliminary Analysis of the Output from a Database Search

When the X!Tandem search is complete, the GPM displays a list of identified protein and peptide sequences (Fig. 5). Almost all investigators will rush to look at the list to see what proteins were in

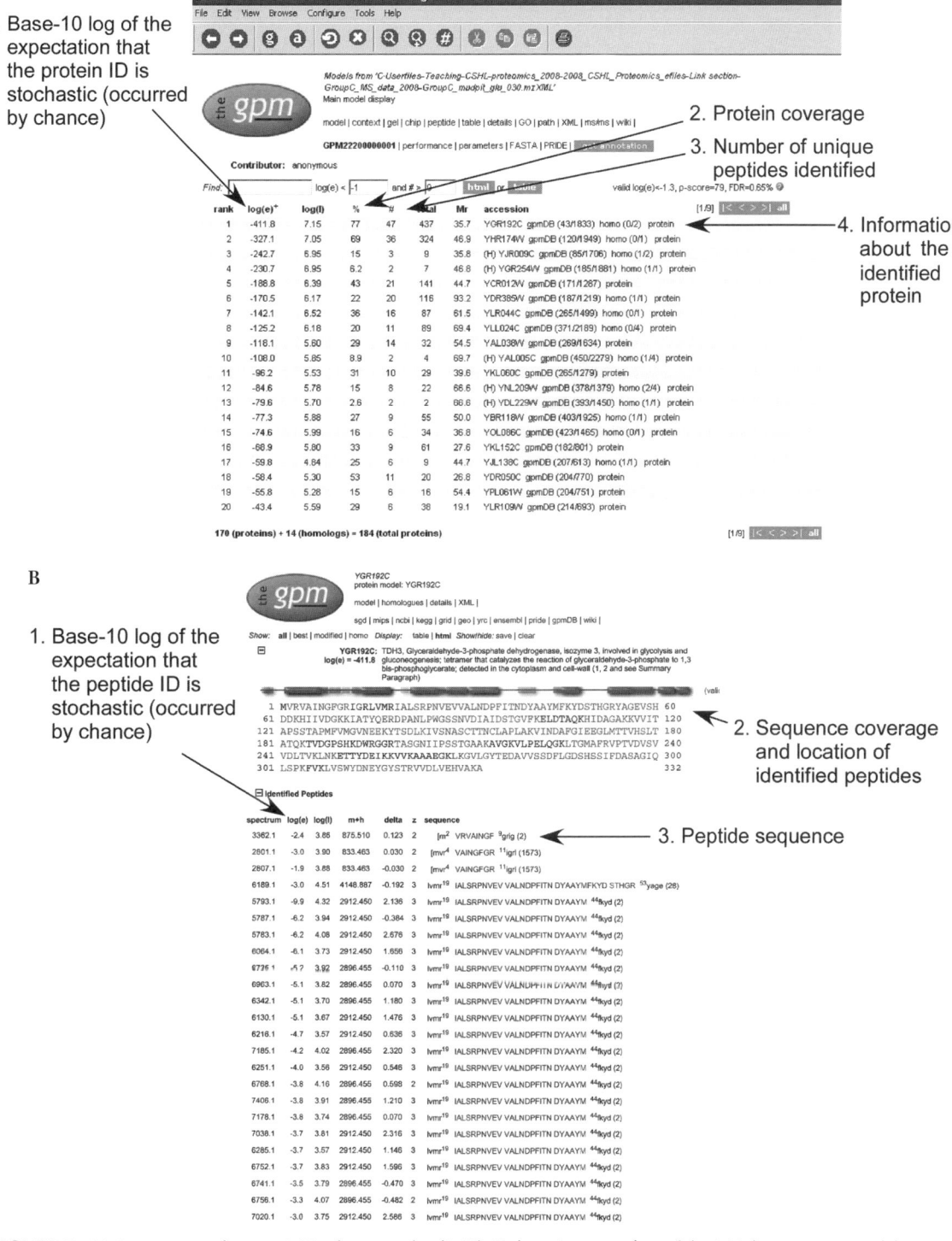

FIGURE 5. (A) Output page from an X!Tandem search of MS/MS data. A screen shot of the initial output or "model" page from X!Tandem showing the key information used to initially evaluate the list of identified proteins. (1.) log(e)+ is the statistical score indicating the significance of the identification. The list of proteins is initially sorted by the proteins with the most significant identifications. For an initial evaluation, proteins with log(e)+ scores below –1 are a good place to begin evaluating the data. (2. and 3.) The percent protein coverage and number of unique peptides identifying the protein are useful metrics for initially evaluating the results. Protein identifications based on ≥ 2 unique peptides are typically considered accurate. (4.) Information about the identified protein includes its accession number and gene name. Other information typically includes its predicted biological function, cellular location, and biological process. (B) Information on the protein identified by X!Tandem. A screen shot of the "protein model" page from X!Tandem illustrating the key information available. The page provides details on the individual spectrum and the peptides identifying a specific protein. The protein coverage section provides a graphical and text view of the location of the identified peptides.

their sample. Often investigators rapidly scan down the list of identified proteins looking for the proteins they consider exciting. These proteins are commonly called the "shiny pebbles." The evidence that these "shiny pebbles" are accurately identified may be marginal. Overall, the interpretation and validation of the list of peptides and proteins is one of the most exciting and challenging parts of proteomics.

Before interpreting the results, it is important to remember several key points.

- The database search is unbiased. All protein and peptide sequences in the database are typically compared with each MS/MS spectrum. The search engine did not make any assumptions about what proteins might be in the sample. This is probably the most important and powerful aspect of the approach. Unexpected proteins are identified that investigators would not have predicted to be in the sample.

- Only peptides and proteins in the protein database will be listed in the output. If a protein or peptide sequence is not in the database, it will not be identified. The search algorithms do not perform de novo sequencing of the MS/MS spectra and derive peptide sequences. The programs only match peptide sequences in a database to the MS/MS spectra using the parameters in the setup file.

- Larger proteins are easier to identify than smaller proteins. The LC-MS/MS approach will only fragment a subset of the total tryptic peptides in the sample. The more tryptic peptides a protein generates, the greater the chance a peptide from the large protein will be selected for fragmentation. Small proteins that only generate a few tryptic peptides could escape detection.

- The LC-MS/MS and MALDI-MS approaches are biased towards fragmenting peptides from abundant proteins. Since the selection process for fragmentation of precursor ions is typically based on ion intensity, the more intense ions will be selected for fragmentation and the less intense ions will be ignored and possibly escape fragmentation. While peptides ionize at different efficiencies, on average, the more abundant a peptide or protein is, the more likely it will generate peptides with strong precursor ion signals. Low-abundance peptides and proteins could escape detection.

- Peptides with unexpected amino acid changes or substitutions will not be matched to a sequence in the database and will not be listed. Once again, protein sequences not in the database will not be identified. Some database search programs will perform what is called "homology" searching, which allows for amino acid substitutions.

- Any peptides with modified amino acids that were not specified in the search parameters will not be listed. These peptides will escape detection. The search programs will only search and list peptides with modified amino acids if they have been included in the search parameters.

Given the strengths and weaknesses of the LC-MS/MS coupled with database search analysis, next we describe an approach for initially evaluating the results.

1. Select criteria for initially accepting database search results as correct.

    A number of studies have examined scoring criteria for deciding whether to accept or reject a peptide identification from the different database search algorithms (Washburn 2001; Keller 2002; MacCoss 2002; Nesvizhskii et al. 2003 ; Peng 2003; Sadygov et al. 2004). While the issue is still debated, two methods have been adopted generally. One method calculates a probability score for a peptide or protein identification and the second method estimates the false discovery rate (FDR) after the database search. For both methods, a value of 0.05 is typically a good starting place for initially accepting a peptide identification.

2. Order the list of identified proteins by the number of peptides and tandem mass spectra identifying a protein in the sample. The higher the number of peptides significantly

matching a protein, the more likely the protein is identified accurately and is not a stochastic hit.

> Since each unique peptide sequence matching a tandem mass spectrum is an independent event, each peptide from a protein that significantly matches the unique tandem mass spectra increases the probability that the protein is accurately identified. As an example, for two independent peptides identifying the same protein with 0.05 probability, the probability the protein is incorrectly identified is 0.05 × 0.05 = 0.0025. In other words, there is a 1 out of 400 chance the protein is incorrectly identified. If only one peptide identifies the protein at 0.05 probability, there is a 1 in 20 chance the protein has been falsely identified. Two independent peptides identifying a protein is strong evidence that the protein is accurately identified. In publishing proteomic data, two independent peptides identifying a protein by tandem mass spectrometry analysis is typically sufficient for the journal to accept the result.

3. Recognize proteins that are identified by unique peptides. In higher eukaryotic organisms, extensive gene duplication and alternative RNA splicing has resulted in a large number of proteins with similar sequences. Protein databases may contain multiple sequences for a protein that represent alternative forms that differ by only a single amino acid. Because peptides are frequently shared by multiple proteins in a database, the assembly of identified peptides into a list of proteins can drastically overstate the number of proteins in samples. It is important to recognize which proteins are identified by unique peptides and which proteins are indistinguishable because only shared peptides are identified. For proteins that only share peptide sequences, the identified proteins should be reported as a protein group.

4. Identify proteins or protein groups that are indistinguishable. For complex eukaryotic organisms (like humans and mice) with rampant alternative splicing, gene duplications, and polymorphisms, protein databases often contain large numbers of proteins with similar sequences. For example, human protein databases may contain many serum albumin and immunoglobin proteins that differ only by a few amino acid residues. These proteins will be indistinguishable because there are no unique peptides allowing the investigator to determine which protein produced the peptides. In these cases, reporting the smallest numbers of proteins necessary to explain the observed peptides (parsimony) is encouraged (Carr et al. 2004; Bradshaw 2005).

5. For proteins or modified peptides identified by a single spectrum, manual evaluation of the spectra and the matching sequence is required (see Protocol 3).

PROTOCOL 2

Using ProteinPilot Software for Peptide and Protein Identification

ProteinPilot is a software package used to identify proteins from peptides generated from 2D gel and MudPIT experiments, including quantification of peptides from the iTRAQ, SILAC, and ICAT techniques (Shilov et al. 2007). It features a novel approach to peptide identification, called the Paragon algorithm (Shilov et al. 2007). Paragon assesses each MS/MS spectrum from a data set to determine which spectra are worth scoring. In other words, it acts as a filter to remove spectra that are less likely to yield reliable peptide identifications. It establishes a threshold for score-worthy spectra based on two inputs: a Sequence Temperature Value (STV) and feature probability. The STV identifies "hot" versus "cold" regions of the database by computing a quantity, based on extracted de novo sequence tags from an MS/MS spectrum, which reflects the degree that each theoretical peptide from a database matches the MS/MS spectrum. Feature probabilities factor in things like posttranslational modifications (PTMs), digestion events (missed or nonspecific cleavages), mass tolerances, and substitutions. A peptide with a "hot" STV receives more search time and is considered for more modifications or other feature events, while "cold" peptides receive less search time and are considered for only the more common PTMs and features. A threshold is then computed, based on STV and feature probabilities, that determines which MS/MS spectra should be scored and which should be discarded. Alternatively, the software provides an interface for searching data with the Mascot algorithm.

While ProteinPilot is an innovative and effective proteomic tool for searching databases, it is currently only compatible with data generated from instruments such as the 4700 MALDI TOF/TOF Analyzer, manufactured by Applied Biosystems. Data can only be uploaded into ProteinPilot via a direct connection to the instrument database. Externally generated data must be converted to mgf format.

This protocol describes procedures for analyzing data generated from a 2D gel experiment (Experiment 1) or data acquired from an iTRAQ experiment (Experiment 7). No file format manipulation is necessary because the software reads the raw file generated by the mass spectrometer directly from a table in the instrument data. This protocol provides procedures for loading data from the database, setting up a search, and exporting and storing the results. An overview of the software features relevant to data output for both 2D gel and iTRAQ data will be covered with an emphasis on data interpretation.

MATERIALS

Computer running Windows XP SP2 and ProteinPilot (version 2.0.1)
Spot set to be analyzed or iTRAQ data

PROCEDURE

Loading Data Files (Spot Sets) in ProteinPilot

1. Open the software by double clicking the ProteinPilot icon on the desktop.

2. In the "Workflow Tasks" task bar, click "Identify Proteins." Figure 6 displays the "Identify Proteins" screen view.

3. Click "Add 4000 Series Data" (Fig. 6A). Select the appropriate spot set(s) to be analyzed. The data set(s) appears in the "Data Sets to Process" window (Fig. 6B).

    For the CSHL course, select the appropriate spot set(s) from the proteomics course project to be analyzed.

Setting Up a Database Search with ProteinPilot for iTRAQ Data or 2D Gel Data

4. In the "Process Using" box, select "Paragon" and click "Edit" (Fig. 6C). The "Paragon Method" window appears (Fig. 7). The last method used appears by default. This can be modified to suit the present application and saved accordingly for future searches. It can then be loaded for future searches through the "Paragon Method" drop-down list shown in Figure 7.

    The option for analyzing data with Mascot is also available but will not be covered here. To use Mascot, the data must be loaded in mgf format.

5. In the "Describe Sample" box, select either "iTRAQ 4 Plex" ("Peptide Labeled") or, for 2D gel data, "Identification" from the "Sample Type" drop-down list.

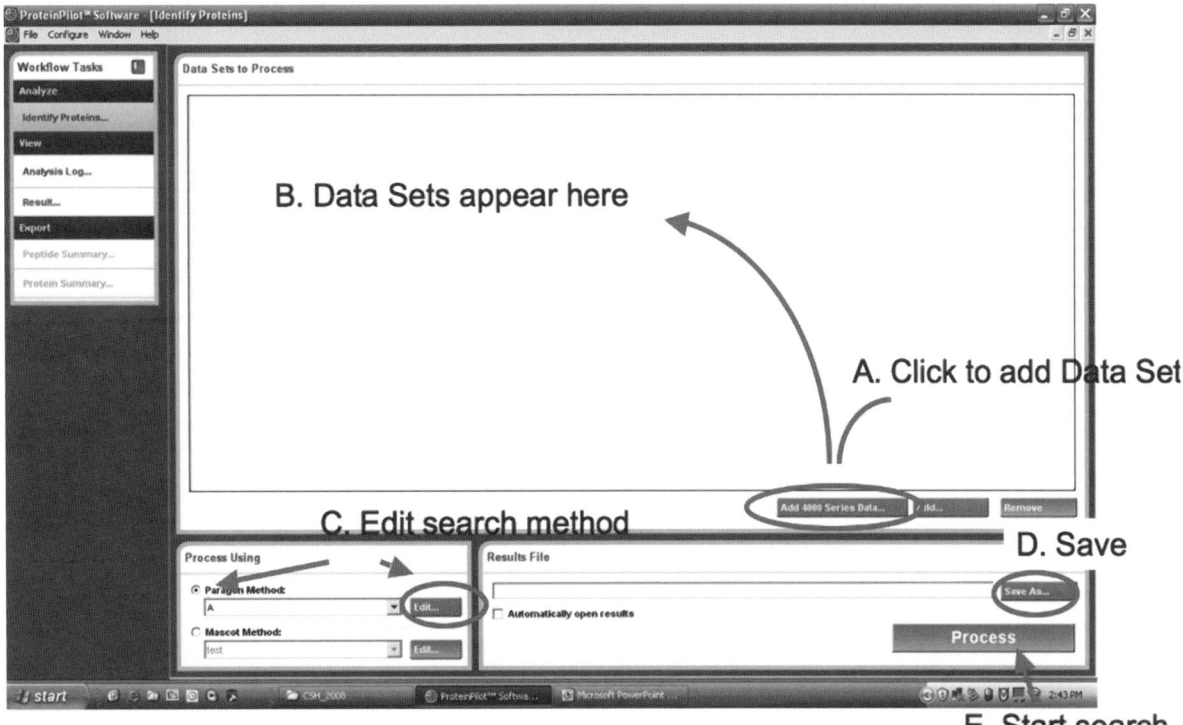

FIGURE 6. The main window of the "Identify Proteins" task in ProteinPilot and workflow for (A,B) loading data for processing, (C) setting search parameters, (D) establishing a directory for saving results, and (E) processing the data.

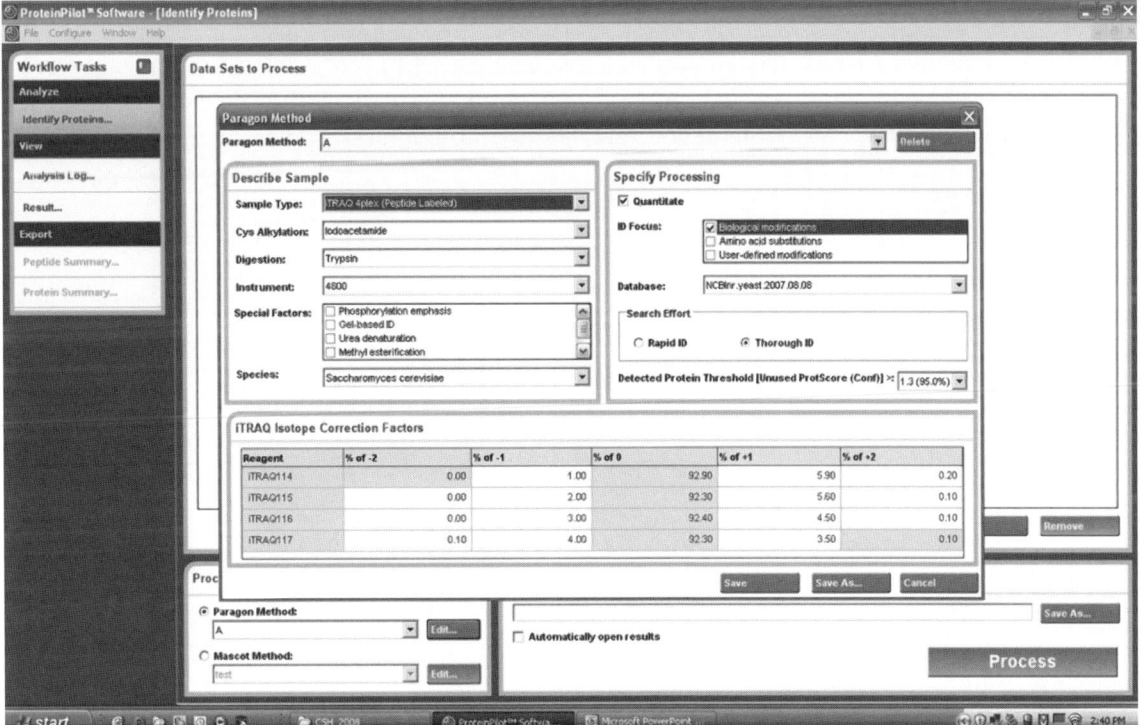

FIGURE 7. The "Paragon Method" editor page.

6. In the "Cys Alkylation" drop-down list, select "Iodoacetamide."
7. In the "Instrument" drop-down list, select "4700."
8. In the "Special Factors" box:
   a. (for iTRAQ data) leave all check boxes unchecked.
   b. (for 2D gel data) select "Gel-based ID."
9. In the "Species" drop-down list, select *"Saccharomyces cerevisiae."*
10. In the "Specify Processing" box:
    a. (for iTRAQ data) make sure the "Quantitate" check box is checked.
    b. (for 2D gel data) make sure the "Quantitate" check box is NOT checked.
11. In the "ID Focus" box, check "biological modifications."
12. In the "Database" drop-down list, select "NCBInr.yeast.2007.08.08."

    Databases in the drop-down list are in FASTA format and located in the directory C:\Applied Biosystems MDS Sciex\ProteinPilot Data\SearchDatabases.

13. In the "Search Effort" box, select "Thorough ID."
14. In the "Detected Protein Threshold" drop-down list, select the desired protein confidence interval threshold. The results will display all protein hits that score above the selected threshold.
15. (for iTRAQ data only) In the "iTRAQ Isotope Correction Factors" box, enter the suggested values as listed in the certificate of analysis that comes with the iTRAQ kit. These values adjust the calculated iTRAQ ratios to compensate for the sub-100% yield in isotopic enrichment of each of the iTRAQ reagents during synthesis.
16. Click "Save As..." to save the search parameters.

    For students in the course, save the parameters by group name and application (e.g., "group A iTRAQ").

17. In the Results File box (Fig. 6D), click "Save As..." to select the results file name and directory. By default, the data will be saved (in .group format) in the directory C:\Applied Biosystems MDS Sciex\ProteinPilot Data\Results, although this can be changed by the user.
18. Click "Process" (Fig. 6E).

Loading Result Files in ProteinPilot

19. In the "Workflow Tasks" task bar, click "Result" (Fig. 6).
20. Select the results file (.group format) from the directory C:\Applied Biosystems MDS Sciex\ProteinPilot Data\Results.

Overview of the Results Screen in ProteinPilot

When a results file is open (Fig. 8), either three or four tabs will be available for data sorting and interpretation. If search parameters were selected for identification only, such as those selected for 2D-gel identifications, then the tabs "Protein ID," "Spectra," and "Summary Statistics" will be available. If quantification parameters were selected in the search, such as those selected for iTRAQ, then the "Protein Quant" tab will also be available.

The "Protein ID" tab (Fig. 8) lists all proteins identified above the confidence interval thresh-

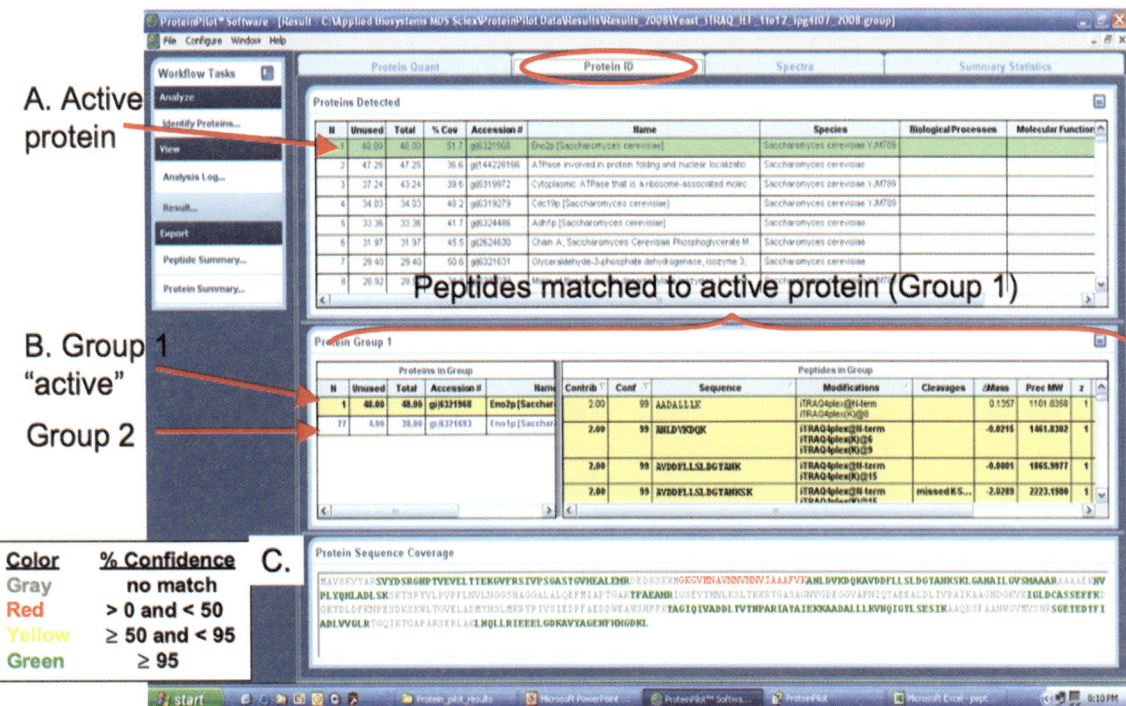

FIGURE 8. The "Protein ID" window of the results file. (A) The active protein is shown in green. (B) All peptide information for the activated protein appears in the "Protein Group" pane. (C) Sequence coverage of the active protein is displayed with color coding (*inset*) in the "Protein Sequence Coverage" pane.

old specified in the search parameters. In the "Proteins Detected" pane, the proteins are listed with scoring information, name, accession number, and species, among other things. Sorting by score, protein name, or any other category is achieved by clicking the column headers. Repetitive clicking on the column header allows for sorting in ascending or descending order. Each protein is assigned two scores: Unused and Total. These are defined below. The active protein in the "Proteins Detected" pane is illuminated green (Fig. 8A). Clicking on a protein row activates it. The "Protein Group" table (Fig. 8B) displays the peptide information that pertains to the activated protein as well as any proteins that have been grouped with it by the Pro Group algorithm. Each peptide has a contrib score which represents the peptide's scoring contribution to the protein's Total Score. The Total Score (third column from the left in the "Protein Group" table) is, therefore, a sum of all contrib scores from peptides assigned to it. However, these are not necessarily unique peptides, so some peptides can be included in the Total Score of more than one protein. Proteins that have peptide hits in common are grouped (Fig. 8B). The Unused Score (second column from the left in the "Protein Group" table) exhibits the peptide scores that have not also been assigned to a higher-scoring protein. It is a reflection of the peptide's uniqueness to the protein. Finally, the "Protein Sequence Coverage" pane (Fig. 8C) displays a view of the sequence coverage of the activated protein. The colors reflect the confidence in the matched peptides assigned to the protein.

The "Spectra" tab lists specific details of all MS/MS spectra that yielded peptide hits. It includes scores, modifications, and a spectral view of the spectrum itself with annotation. If the data were acquired and searched for iTRAQ quantification, then iTRAQ ratios are also included. It also lists other peptides that matched the spectrum with lower confidence. The "Summary Statistics" tab provides details about the number of proteins that were identified at different confidence intervals, including the confidence interval selected for viewing when the search parameters were set. It also summarizes the results parameters, analysis parameters, and quantification settings that were used to search and analyze the data.

The "Protein Quant" tab displays relative quantification of all confidently identified proteins and only appears if the appropriate settings for quantification were entered in the search parameters. The "Protein Quant" tab (Fig. 9) is arranged similarly to the "Protein ID" tab except that it focuses more on quantification and less on identification. In the "Proteins Detected" pane, proteins are listed with essentially the same information found in the "Protein ID" tab, with the inclusion of iTRAQ ratios. Proteins are activated by clicking on the appropriate row, illuminating it green (Fig. 9A). The user controls what information from the "Protein ID" or "Protein Quant" tab to display. By right clicking any of the column headings, a checkbox appears allowing the user to show or hide information (Fig. 9B). (This feature is available for every table in the software.) The peptide information in the "Peptide Quantitation" pane corresponds to the active protein (Fig. 9C). The "Peptide Quantitation" pane displays information about all peptides assigned to the active protein with emphasis on the iTRAQ ratios measured from the peptide. If the checkbox in the "Used" column is checked, then the iTRAQ ratios for the peptide were used in calculating the overall iTRAQ ratios for the active protein. This provides user flexibility in deciding to use the data or omit it. The "Annotation" column specifies whether the decision to use, or not to use, the quantification data from the peptide was determined by the user or by the software. If it displays "auto," the current status in the "User" column was selected by the software. If it displays "manual," the current status in the "User" column was selected by the user. Mass spectral views of the active peptide iTRAQ reporter ion m/z region and precursor m/z region are displayed in the bottom pane (Fig. 9D).

The most important information in the "Protein Quant" tab is the determination of which proteins exhibit changes in abundance across samples. It is therefore practical to sort the protein infor-

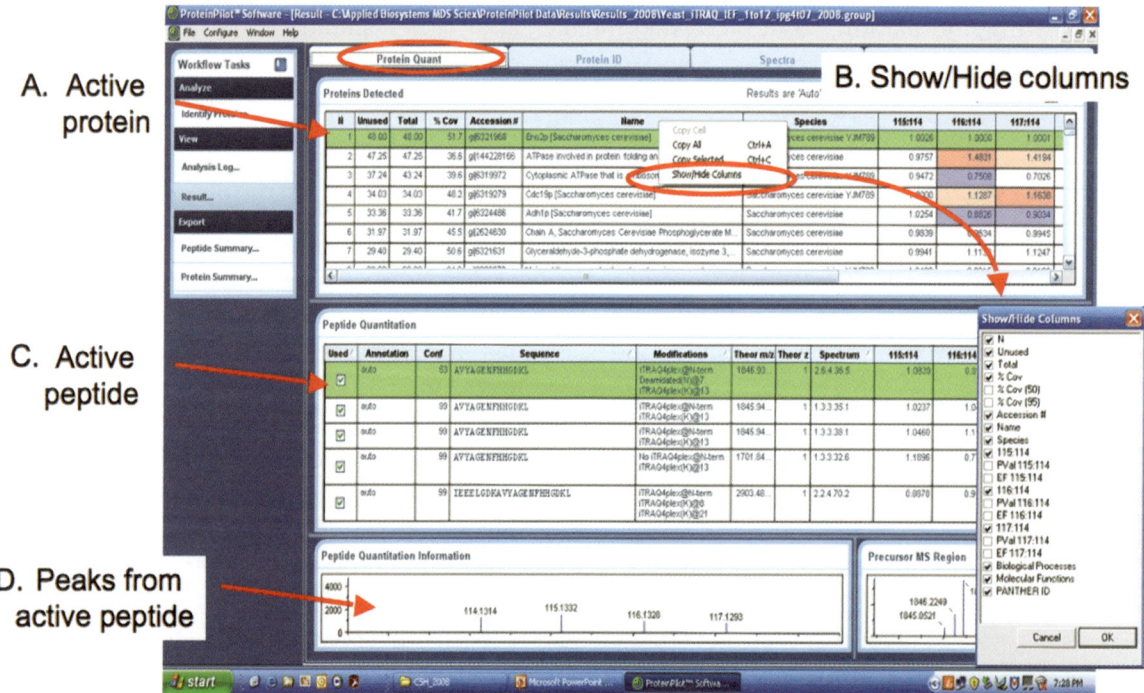

FIGURE 9. The "Protein Quant" window of the results file. (A) The active protein is shown in green. (B) All peptide quantification information for the activated protein appears in the "Protein Group" pane. (C) Mass spectral views of the iTRAQ reporter ion and precursor m/z regions are displayed. (D) Columns in any table can be added or hidden with the "Show/Hide Columns" feature.

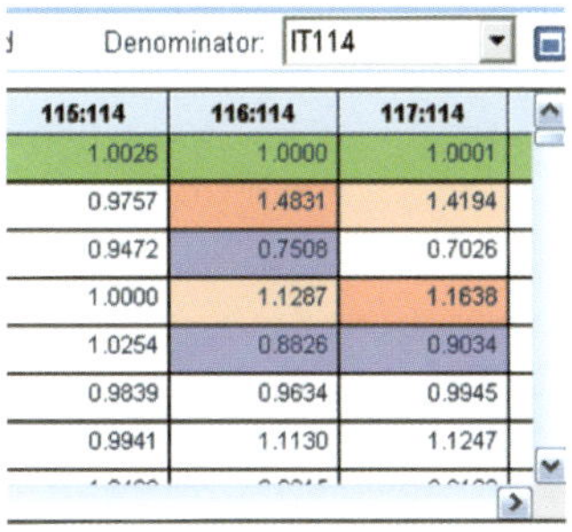

Color	P-value	Ratio
Dark red	< 0.001	> 1
Medium red	0.001 to < 0.01	> 1
Light red	0.01 to < 0.05	> 1
No color	>= 0.05	Any
Light blue	0.01 to < 0.05	< 1
Medium blue	0.001 to < 0.01	< 1
Dark blue	< 0.001	< 1

FIGURE 10. Color coding scheme representing the *p*-value of the iTRAQ ratios.

mation by iTRAQ ratios. By clicking one of the iTRAQ ratio columns in the "Proteins Detected" pane, the protein table automatically sorts by that column in ascending order. Clicking the column heading again will sort the data by that column in descending order. This provides a quick view for which proteins represent targets for further validation. The iTRAQ ratios in the "Proteins Detected" pane appear with different colored backgrounds. Colors reflect the precision, in the form of a *p*-value, by which the ratio was determined from each protein's peptide hits. The color coding is explained in Figure 10.

PROTOCOL 3

Evaluating an MS/MS Spectrum that Matches a Peptide Sequence from a Database Search Program

In the course, we train students using this protocol to decide whether to accept or reject a peptide sequence that matches an MS/MS spectrum after a database search. In the search research, there are always both true and false identifications due to the random or stochastic matching between experimental and theoretical data. When tandem mass spectra contain limited fragmentation information, the database search engines may lead to incorrect identifications. Especially for proteins or posttranslational modification identified from a single MS/MS spectrum, it is necessary to validate identifications by vigilant inspection of the spectra.

1. Evaluate the MS/MS spectrum quality.

   A strong criterion for spectral quality is the presence of background noise (see Fig. 2A). No background noise and only ions of similar intensity coming up from a flat baseline is an indication that the precursor ion was not a peptide. A spectrum with few ions or low total ion intensity is an indication of a peptide in low abundance and may contain inadequate information to validate the peptide hit.

2. Evaluate how extensively the peptide's predicted b- and y-ions match the product ions in the spectrum (Fig. 11).

   Most of the major product ions in the spectrum should match either the b- or y-ions from the sequence. A large number of unexplained intense peaks throughout the spectrum is an indication of an incorrect identification. Examine the continuity of the b- or y-ion series. One should see a continuous string of 3 to 4 ions of the same series as opposed to isolated ions here and there. Complementing b- and y-ion pairs are a good indication of a correct identification.

   While tryptic peptides typically generate doubly and triply charged precursor ions during ESI-LC-MS/MS, small tryptic peptides (< 6–8 residues) will be seen commonly as singly charged. Importantly, the short peptides will generate a limited number of fragment ions. The small num-

A

b ions	b1	b2	b1-17	b2-8.5	b1-18	b2-9	y ions	y1	y2	y1-17	y2-8.5	y1-18	y2-9
b1 V	100.1326	50.5663	83.1326	42.0663	82.1326	41.5663	y1 K	147.1742	74.0871	130.1742	65.5871	129.1742	65.0871
b2 P	197.2493	99.12465	180.2493	90.62465	179.2493	90.12465	y2 V	246.3068	123.6534	229.3068	115.1534	228.3068	114.6534
b3 T	298.3544	149.6772	281.3544	141.1772	280.3544	140.6772	y3 T	347.4119	174.20595	330.4119	165.70595	329.4119	165.20595
b4 V	397.487	199.2435	380.487	190.7435	379.487	190.2435	y4 L	460.5714	230.7857	443.5714	222.2857	442.5714	221.7857
b5 D	512.5756	256.7878	495.5756	248.2878	494.5756	247.7878	y5 D	575.66	288.33	558.66	279.83	557.66	279.33
b6 V	611.7082	306.3541	594.7082	297.8541	593.7082	297.3541	y6 V	674.7926	337.8963	657.7926	329.3963	656.7926	328.8963
b7 S	698.7864	349.8932	681.7864	341.3932	680.7864	340.8932	y7 V	773.9252	387.4626	756.9252	378.9626	755.9252	378.4626
b8 V	797.919	399.4595	780.919	390.9595	779.919	390.4595	y8 S	861.0034	431.0017	844.0034	422.5017	843.0034	422.0017
b9 V	897.0516	449.0258	880.0516	440.5258	879.0516	440.0258	y9 V	960.136	480.568	943.136	472.068	942.136	471.568
b10 D	1012.1402	506.5701	995.1402	498.0701	994.1402	497.5701	y10 D	1075.2246	538.1123	1058.2246	529.6123	1057.2246	529.1123
b11 L	1125.2997	563.14985	1108.2997	554.64985	1107.2997	554.14985	y11 V	1174.3572	587.6786	1157.3572	579.1786	1156.3572	578.6786
b12 T	1226.4048	613.7024	1209.4048	605.2024	1208.4048	604.7024	y12 T	1275.4623	638.23115	1258.4623	629.73115	1257.4623	629.23115
b13 V	1325.5374	663.2687	1308.5374	654.7687	1307.5374	654.2687	y13 P	1372.579	686.7895	1355.579	678.2895	1354.579	677.7895

B

b ions	b1	b2	b1-17	b2-8.5	b1-18	b2-9	y ions	y1	y2	y1-17	y2-8.5	y1-18	y2-9
b1 P	98.1167	49.55835	81.1167	41.05835	80.1167	40.55835	y1 N	133.1039	67.05195	116.1039	58.55195	115.1039	58.05195
b2 V	197.2493	99.12465	180.2493	90.62465	179.2493	90.12465	y2 N	247.2078	124.1039	230.2078	115.6039	229.2078	115.1039
b3 T	298.3544	149.6772	281.3544	141.1772	280.3544	140.6772	y3 T	348.3129	174.65645	331.3129	166.15645	330.3129	165.65645
b4 V	397.487	199.2435	380.487	190.7435	379.487	190.2435	y4 L	461.4724	231.2362	444.4724	222.7362	443.4724	222.2362
b5 E	526.6025	263.80125	509.6025	255.30125	508.6025	254.80125	y5 N	575.5763	288.28815	558.5763	279.78815	557.5763	279.28815
b6 T	627.7076	314.3538	610.7076	305.8538	609.7076	305.3538	y6 G	632.6283	316.81415	615.6283	308.31415	614.6283	307.81415
b7 V	726.8402	363.9201	709.8402	355.4201	708.8402	354.9201	y7 I	745.7878	373.3939	728.7878	364.8939	727.7878	364.3939
b8 I	839.9997	420.49985	822.9997	411.99985	821.9997	411.49985	y8 V	844.9204	422.9602	827.9204	414.4602	826.9204	413.9602
b9 G	897.0517	449.02585	880.0517	440.52585	879.0517	440.02585	y9 T	946.0255	473.51275	929.0255	465.01275	928.0255	464.51275
b10 N	1011.1556	506.0778	994.1556	497.5778	993.1556	497.0778	y10 E	1075.141	538.0705	1058.141	529.5705	1057.141	529.0705
b11 L	1124.3151	562.65755	1107.3151	554.15755	1106.3151	553.65755	y11 V	1174.2736	587.6368	1157.2736	579.1368	1156.2736	578.6368
b12 T	1225.4202	613.2101	1208.4202	604.7101	1207.4202	604.2101	y12 T	1275.3787	638.18935	1258.3787	629.68935	1257.3787	629.18935
b13 N	1339.5241	670.26205	1322.5241	661.76205	1321.5241	661.26205	y13 V	1374.5113	687.75565	1357.5113	679.25565	1356.5113	678.75565

FIGURE 11. Evaluating the first and second peptide hits to an MS/MS spectrum. Shown are the top two peptide hits for an MS/MS spectrum after a Sequest search of the acquired data using a yeast protein database. (A) Highest-scoring peptide hit (VPTVDVSVVDLTVK) from the yeast Tdh1/2/3 protein (Cn = 4.6). (B) Second-best scoring peptide hit (PVTVETVIGNLTNN) from the yeast Meu1 protein (Cn = 2.0). Overall, the first peptide hit in Panel A is considered accurate because of several important reasons. First, it is a good spectrum because of the overall ion intensity and the presence of strong ion signals above the background noise. Second, the majority of the major fragment ions are labeled as a b- or y-ion. Third, the peptide hit is a canonical tryptic fragment. Fourth, in the table showing experimental ion data matching the theoretical fragmentation ion values, the peptide has a long, continuous string of b- and y-ions. Finally, the b- and y-ion pairs are complementary (e.g., the sum of the complementary b-y ion pairs equals the precursor ion mass). The second hit in Panel B is considered erroneous for a number of reasons. Many of the major ions are unlabeled. The second-best hit is a noncanonical tryptic peptide. The table below the spectrum shows a random matching of experimental data to the theoretical values with no recognizable pattern.

ber of fragment ions hinders the search algorithm's ability to identify the correct peptide and the user's ability to validate the sequence.

It is often not possible to explain all the fragment ions that are observed in a spectrum. However, for doubly and triply charged tryptic peptides, the majority of the most abundant peaks in the *m/z* range above the precursor ion should be evidence of a continuous y-ion series.

Tandem mass spectra that are extremely complex may be caused by fragmentation of two different peptides simultaneously. This can be determined by examining the precursor scan for evidence of two precursor ions. After identifying one peptide, the unidentified peaks in the MS/MS scan can be used for a second database search.

The moderate resolution of ion trap mass spectrometers often limits the ability to directly determine the charge state of the precursor ions. When the charge state of the precursor ions is unknown, database search algorithms will commonly generate multiple charge states (e.g., +2, +3, +4) for the precursor ions. Each precursor charge state is searched separately against the protein database and multiple peptide sequences are reported. It is important for users to accept only one peptide identification.

It is important to recognize that the CID conditions used to fragment precursor ions are different for ion trap and TOF/TOF analyzers. Ion trap instruments fragment precursor peptide ions by inducing low-energy collisions with an inert gas, typically helium, within the ion trap analyzer. Once the precursor ion fragments, the resulting fragment ions do not fragment again. In ion traps, the mechanism of CID does not allow for trapping of fragment ions below 28% of the precursor mass. This "1/3 rule" or "low mass cut-off" is evidenced by a lack of ions in the low mass range of the tandem spectra. In the TOF/TOF instrument, on the other hand, fragmentation is variable. The TOF/TOF mass spectrometer is capable of high-energy collisions that can fragment side chain bonds and low-energy collisions that fragment primary peptide bonds (Khatun et al. 2007). Internal fragment ions and y-ions are commonly observed. For users of QTOF and triple quadrupole instruments, CID occurs in a separate radio-frequency collision cell so all of the ions entering the cell are excited, and secondary fragmentation of b- and y-ions may occur. Compared to the ion trap, the TOF/TOF, QTOF, triple quadrupole, and FTICR mass spectrometers all retain the low mass product ions. Finally, while the precursor ions for tryptic peptides using ESI are typically doubly or triply charged, the precursor ions from a MALDI source are generally singly charged ions. For all these reasons, the MS/MS spectra generated from an ion trap, TOF/TOF, QTOF, and other mass analyzers for the same peptide sequence will not be identical and can show substantial differences.

3. Evaluate the peptide sequence and spectrum for unique sequence effects on intensity (Fig. 12).

Fragmentation tends to occur in the middle of the peptide rather than near the termini. In ion trap mass spectrometers, y-ion intensity is typically twice as strong as b-ion intensity (Tabb et al. 2003, 2006). Specific residues tend to have unique effects on fragmentation. Recognizing these effects on the spectrum increases the confidence of the identification. Most notably, peptides with an internal proline have a strong tendency to fragment on the amino-terminal side of the proline residue (Fig. 12). This is commonly referred to as the "proline effect." Aspartic acid residues have a strong tendency to fragment on their carboxy-terminal side (Kapp et al. 2003). Studies of the frequency of peptide fragmentation based on amino acid pairs shows that isoleucine, valine, and leucine residues favor fragmentation on their carboxy-terminal side and glycine and serine favor fragmentation on their amino-terminal sides (Tabb et al. 2003). Peptide bonds between asparagine and glycine are very labile (Kapp et al. 2003).

4. Examine the spectrum for neutral-loss product ions. Both precursor and fragment ions may lose small neutral molecules (Fig. 13).

Neutral losses from the precursor can cause major peaks in the fragmentation spectrum at a lower *m/z* value compared to the precursor value. Most notably, phosphopeptides containing phosphoserine or phosphothreonine residues will readily lose phosphoric acid (–98 Da) when using low-energy CID, such as an ion trap (Schlosser et al. 2001). This observation is extremely useful for recognizing phosphopeptides (Fig. 13). For doubly charged phosphopeptide precursor ions, the most intense ion in the fragmentation spectrum will be the neutral-loss ion (–49 Da from the precur-

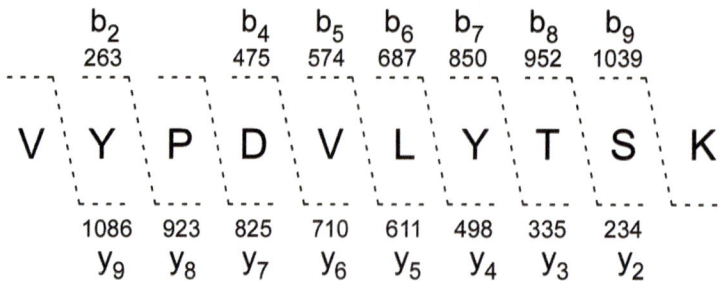

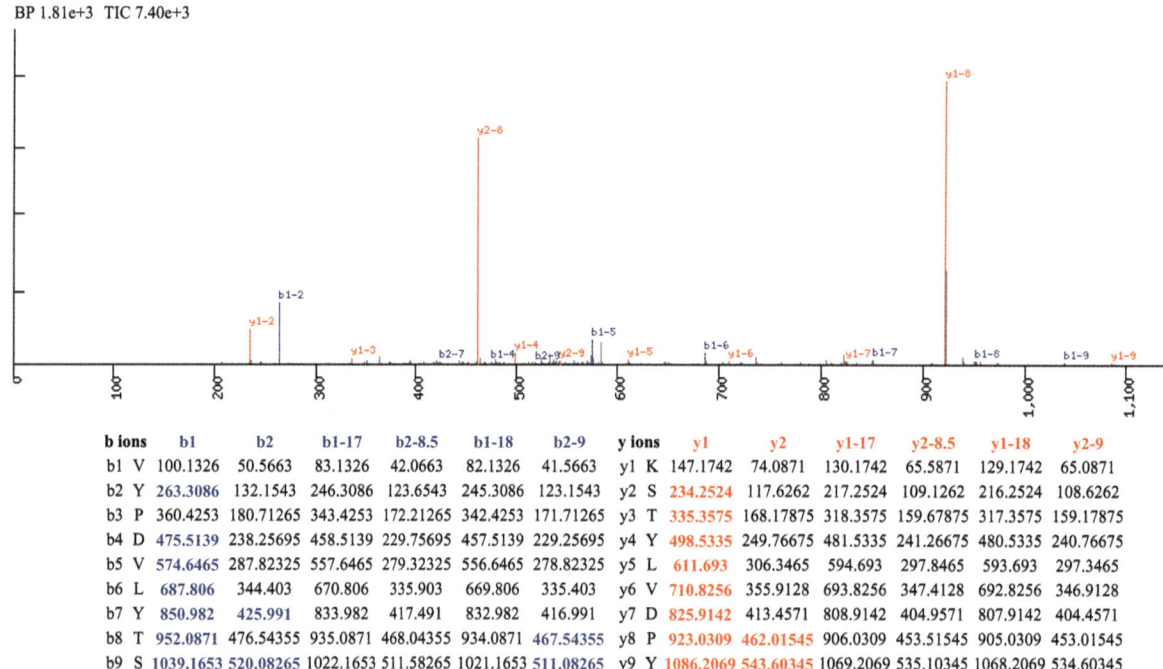

FIGURE 12. Peptide with a proline residue generates intense fragment ions. The MS/MS fragmentation spectrum of the precursor ion (m/z 592.92) shows two intense ions. The Sequest algorithm identified the peptide "VYPDVLYTSK" from the yeast Gpm1 protein as the highest scoring hit (Cn 2.2). The two dominant ions are examples of proline peaks illustrating the intense fragment ions formed on the amino-terminal side of proline residues. The y_{1-8} and y_{2-8} are the singly and doubly charged y_8 ions, respectively. The precursor ion was doubly charged. The table below the labeled spectrum shows the observed fragment ions matching the theoretical fragment ions from the peptide sequence. The b2 and y2 columns are the doubly charged b-and y-ion values. The X–17 and –18 columns (where X is b1 or y1) are the singly charged values for the respective b- and y-ions with neutral losses of ammonia and water, respectively. The X–8.5 and X–9 columns (where X is b2 or y2) are the doubly charged values for the respective b- and y-ions with neutral losses of ammonia and water, respectively.

sor). For triply charged ion phosphopeptides, the most intense ion will be the neutral-loss ion (–32.6 Da from the precursor). In an ion trap's fragmentation spectrum of the phosphorylated peptide, the signal from the b- and y-ions is dramatically reduced relative to the neutral-loss peak. The reduced intensity of the b- and y-ion peaks renders a confident identification more difficult. In particular, it is difficult to identify the exact amino acid that is modified if the peptide contains multiple Ser or Thr residues.

For fragment ions that have a neutral loss, pairs of peaks are typically produced. The neutral-loss ion will typically be 10–20% of the intensity of the intact fragment ion. Fragment b- and y-ions that contain Ser, Thr, Glu, or Asp may lose H_2O (–18 Da). b- and y-ions that contain Asn, Gln, Arg, or Lys may lose ammonia (–17 Da). Precursor ions with Gln at their amino termini can readily lose ammonia (–17 Da), generating a neutral loss peak and a predominant b-17 ion series. Precursor ions with oxidized methionines may lose methane sulfenic acid (–64 Da) (Reid et al. 2004).

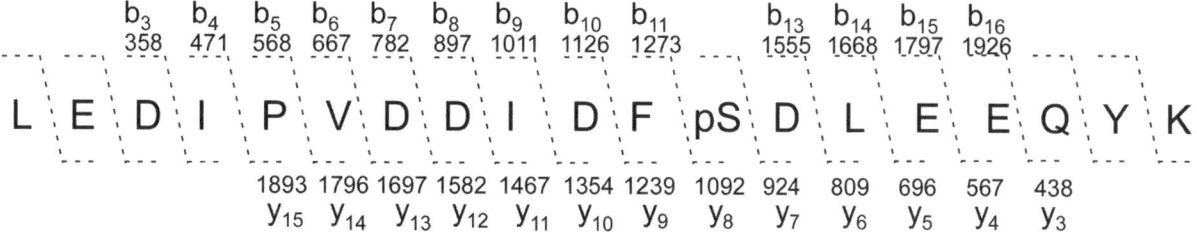

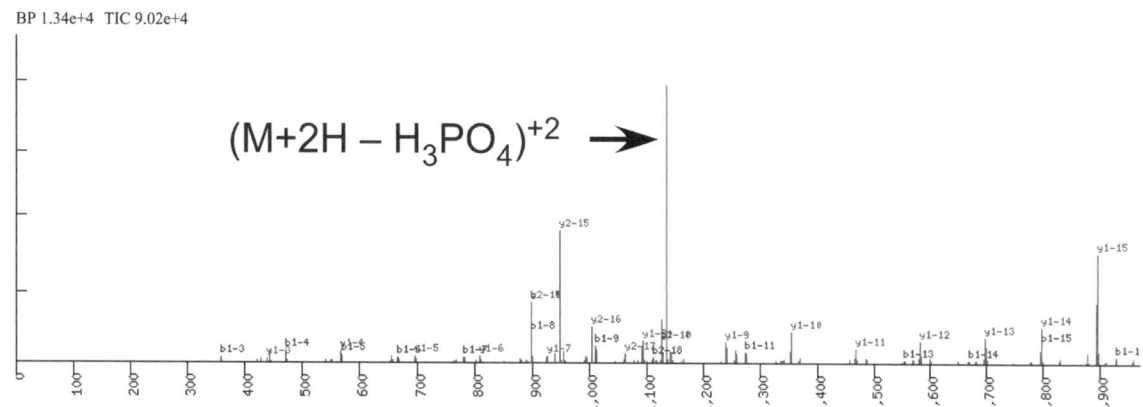

FIGURE 13. MS/MS spectrum of a phosphopeptide. Phosphopeptides can show distinct neutral losses of phosphoric acid (–98 Da) from the precursor and fragment ions. During low-energy CID, especially in ion trap mass spectrometers, phosphopeptides with phosphoserine or phosphothreonine residues readily lose phosphoric acid generating an intense neutral ion in the MS/MS spectrum. In this experiment, to identify phosphorylated peptides, IMAC-Fe^{+3} was used to enrich for phosphopeptides from yeast. The captured peptides were analyzed by LC-MS/MS using an ion trap. This product spectrum was acquired on a 1184 m/z precursor ion. The Sequest search identified the phosphopeptide LEDIPVDDIDFS*DLEEQYK as the top hit (Cn = 5.4). Manual analysis of the MS/MS spectrum identified the most intense ion as a neutral loss of –49 (phosphoric acid) from the doubly charged precursor ion. For singly or triply charged phosphopeptide precursor ions, neutral losses of –98 and –32.6 would have been observed, respectively. In this spectrum, a relatively intense proline peak is also observed.

5. Look for independent spectra identifying the same peptide.

   For abundant peptides, different charge states (e.g., +2 and +3) of the identical peptide may be selected for fragmentation. Since the searches for different charge states are independent of each other, the probability that the identification is correct increases if the searches return the same peptide sequence.

6. Evaluate the basicity of the fragment ions.

   Several studies have examined the influence of basic residues on fragment ion intensities (Paizs and Suhai 2002; Tabb et al. 2004). When a tryptic peptide contains a single basic amino acid residue fragmented by CID (Lys or Arg), the product ions that retain the basic residue are generally more intense compared to the other fragment ion. For fragmentation in ion trap mass spectrometers, the y-ions are typically more intense than the b-ions. When triply charged peptide ions are fragmented by CID, the product ions that contain multiple basic residues (Lys, Arg, His) are more likely to be doubly charged (Tabb et al. 2006). For singly charged precursor ions, one of the fragment ion series may dominate over the other series, especially if the terminal residue is a strongly basic Arg. When validating the MS/MS spectrum, b- and y-ion products that follow these observations support the identification.

7. Examine the low mass ions in the MS/MS spectrum.

   In ion trap mass spectrometers, the mechanism of CID does not allow for trapping of fragment masses below 28% of the precursor mass. As described earlier, ion trap mass spectrometers suffer from the "1/3 rule" during fragmentation and fail to retain the low mass ions. However, the

TOF/TOF and QTOF instruments retain the low mass product ions. The low mass ions may reveal information about the sequence composition of the peptides that can be very useful for validating the peptide hits from the database search. Immonium ions are internal product ions produced as a secondary fragmentation of the amide bond during CID. Their structure is represented by $RCH=H_2N^+$, where R indicates the amino acid side chain. For an amino acid, its immonium ion is 27 Da less than its residue mass. Each amino acid in a peptide has a characteristic immonium ion. The presence of immonium ions in the low mass region of an MS/MS spectrum can indicate the presence of that amino acid in the peptide. The Tyr (136), His (110), Met (104), Pro (70), Phe (120), Trp (159), Leu/Ile (86), and Val (72) immonium ions are most often observed. The sequence composition of the peptide can be verified using the immonium ions. The carboxy-terminal residue of the peptide can be checked against the product ions in the low-mass region of the spectrum. For peptides with a carboxy-terminal Lys residue, the appearance of an ion at 147 may be detected. If the carboxyl terminus is an Arg, a product ion at 175 may be detected. While b_1 ions are rarely seen and y_2 ions are often low intensity, an intense ion pair in the lower m/z range of the MS/MS spectrum separated by 28 Da is frequently observed. The ions correspond to the a_2 and b_2 fragment ions and is the result of the facile loss of CO from the b_2 ion. This pair of ions is commonly called the "a_2/b_2 pair". Appendix 6 lists the m/z values of all the possible b_2-ion combinations of amino acid residue masses.

For cases where accurate identification of the peptide sequence is essential for future experiments, the peptide is often chemically synthesized and the fragmentation pattern of the synthetic peptide is compared to the MS/MS fragmentation of the native precursor ion. The two MS/MS spectrum should show identical fragment ions (m/z) and relative intensities.

REFERENCES

Bradshaw R.A. 2005. Revised draft guidelines for proteomic data publication. *Mol. Cell. Proteomics* **4**: 1223–1225.

Carr S., Aebersold R., Baldwin M., Burlingame A., Clauser K., and Nesvizhskii A.; Working Group On Publication Guidelines For Peptide And Protein Identification Data. 2004. The need for guidelines in publication of peptide and protein identification data. *Mol. Cell. Proteomics* **3**: 531–533.

Chen Y., Kwon S.W., Kim S.C., and Zhao Y. 2005. Integrated approach for manual evaluation of peptides identified by searching protein sequence databases with tandem mass spectra. *J. Proteome Res.* **4**: 998–1005.

Craig R. and Beavis R.C. 2003. A method for reducing the time required to match protein sequences with tandem mass spectra. *Rapid Commun. Mass Spectrom.* **17**: 2310–2316.

Craig R. and Beavis R.C. 2004. TANDEM: Matching proteins with tandem mass spectra. *Bioinformatics* **20**: 1466–1467.

Creasy D.M. and Cottrell J.S. 2004. Unimod: Protein modifications for mass spectrometry. *Proteomics* **4**: 1534–1536.

Dongré A.R., Jones J.L., Somogyi A., and Wysocki V.H. 1996. Influence of peptide composition, gas-phase basicity, and chemical modification on fragmentation efficiency: Evidence for the mobile proton model. *J. Am. Chem. Soc.* **118**: 8365–8374.

Deutsch E. 2008. mzML: A single, unifying data format for mass spectrometer output. *Proteomics* **8**: 2776–2777.

Eng J.K., McCormack A.L., and Yates J.R. 1994. An approach to correlate tandem mass spectral data of peptides with amino acid sequences. *J. Am. Soc. Mass Spectrom.* **5**: 976–989.

Fenyo D. and Beavis R.C. 2003. A method for assessing the statistical significance of mass spectrometry-based protein identifications using general scoring schemes. *Anal. Chem.* **75**: 768–774.

Geer L.Y., Markey S.P., Kowalak J.A., Wagner L., Xu M., Maynard D.M., Yang X., Shi W., and Bryant S.H. 2004. Open mass spectrometry search algorithm. *J. Proteome Res.* **3**: 958–964.

Han D.K., Eng J., Zhou H., and Aebersold R. 2001. Quantitative profiling of differentiation-induced microsomal proteins using isotope-coded affinity tags and mass spectrometry. *Nat. Biotechnol.* **19**: 946–951.

Kapp E.A., Schütz F., Reid G.E., Eddes J.S., Moritz R.L., O'Hair R.A., Speed T.P., and Simpson R.J. 2003. Mining a tandem mass spectrometry database to determine the trends and global factors influencing peptide fragmentation. *Anal. Chem.* **75**: 6251–6264.

Keller A., Nesvizhskii A.I., Kolker E., and Aebersold R. 2002. Empirical statistical model to estimate the accuracy of peptide identifications made by MS/MS and database search. *Anal. Chem.* **74:** 5383–5392.

Khatun J., Ramkissoon K., and Giddings M.C. 2007. Fragmentation characteristics of collision-induced dissociation in MALDI TOF/TOF mass spectrometry. *Anal. Chem.* **79:** 3032–3040.

Link A.J., Eng J., Schieltz D.M., Carmack E., Mize G.J., Morris D.R., Garvin B.M., and Yates J.R. III. 1999. Direct analysis of protein complexes using mass spectrometry. *Nat. Biotechnol.* **17:** 676–682.

MacCoss M.J., Wu C.C., and Yates J.R. III. 2002. Probability-based validation of protein identifications using a modified SEQUEST algorithm. *Anal. Chem.* **74:** 5593–5599.

McCormack A.L., Schieltz D.M., Goode B., Yang S., Barnes G., Drubin D., and Yates J.R. III. 1997. Direct analysis and identification of proteins in mixtures by LC/MS/MS and database searching at the low-femtomole level. *Anal. Chem.* **69:** 767–776.

Nesvizhskii A.I., Keller A., Kolker E., and Aebersold R. 2003. A statistical model for identifying proteins by tandem mass spectrometry. *Anal. Chem.* **75:** 4646–4658.

Paizs B. and Suhai S. 2002. Towards understanding some ion intensity relationships for the tandem mass spectra of protonated peptides. *Rapid Commun. Mass Spectrom.* **16:** 1699–1702.

Paizs B.and Suhai S. 2005. Fragmentation pathways of protonated peptides. *Mass Spectrom. Rev.* **24:** 508–548.

Pappin D.J., Hojrup P. and Bleasby A.J. 1993. Rapid identification of proteins by peptide-mass fingerprinting. *Curr. Biol.* **3:** 327–332.

Peng J., Elias J.E., Thoreen C.C., Licklider L.J., and Gygi S.P. 2003. Evaluation of multidimensional chromatography coupled with tandem mass spectrometry (LC/LC-MS/MS) for large-scale protein analysis: The yeast proteome. *J. Proteome Res.* **2:** 43–50.

Perkins D.N., Pappin D.J., Creasy D.M, and Cottrell J.S. 1999. Probability-based protein identification by searching sequence databases using mass spectrometry data. *Electrophoresis* **20:** 3551–3567.

Reid G.E., Roberts K.D., Kapp E.A., and Simpson R.I. 2004. Statistical and mechanistic approaches to understanding the gas-phase fragmentation behavior of methionine sulfoxide containing peptides. *J. Proteome Res.* **3:** 751–759.

Resing K.A., Meyer-Arendt K., Mendoza A.M., Aveline-Wolf L.D., Jonscher K.R., Pierce K.G., Old W.M., Cheung H.T., Russell S., Wattawa J.L., et al. 2004. Improving reproducibility and sensitivity in identifying human proteins by shotgun proteomics. *Anal. Chem.* **76:** 3556–3568.

Sadygov R.G., Cociorva D., and Yates J.R. III. 2004. Large-scale database searching using tandem mass spectra: Looking up the answer in the back of the book. *Nat. Methods* **1:** 195–202.

Sadygov R.G., Liu H., and Yates J.R. III. 2004. Statistical models for protein validation using tandem mass spectral data and protein amino acid sequence databases. *Anal. Chem.* **76:** 1664–1671.

Schlosser A., Pipkorn R., Bossemeyer D., and Lehmann W.D. 2001. Analysis of protein phosphorylation by a combination of elastase digestion and neutral loss tandem mass spectrometry. *Anal. Chem.* **73:** 170–176.

Shilov I.V, Seymour S.L., Patel A.A., Loboda A., Tang W.H., Keating S.P., Hunter H.L., Nuwaysir L.M., and Schaeffer D.A. 2007. The Paragon algorithm, a next generation search engine that uses sequence temperature values and feature probabilities to identify peptides from tandem mass spectra. *Mol. Cell. Proteomics* **6:** 1638–1655.

Tabb D.L., Fernando C.G., and Chambers M.C. 2007. MyriMatch: Highly accurate tandem mass spectral peptide identification by multivariate hypergeometric analysis. *J. Proteome Res.* **6:** 654–661.

Tabb D.L., Friedman D.B., and Ham A.J. 2006. Verification of automated peptide identifications from proteomic tandem mass spectra. *Nat. Protoc.* **1:** 2213–2222.

Tabb D.L., Huang Y., Wysocki V.H., and Yates J.R. III. 2004. Influence of basic residue content on fragment ion peak intensities in low-energy collision-induced dissociation spectra of peptides. *Anal. Chem.* **76:** 1243–1248.

Tabb D.L., McDonald W.H., and Yates J.R. III. 2002. DTASelect and Contrast: Tools for assembling and comparing protein identifications from shotgun proteomics. *J. Proteome Res.* **1:** 21–26.

Tabb D.L., Smith L.L., Breci L.A., Wysocki V.H., Lin D., and Yates J.R. III. 2003. Statistical characterization of ion trap tandem mass spectra from doubly charged tryptic peptides. *Anal. Chem.* **75:** 1155–1163.

Washburn M.P., Wolters D., and Yates J.R. III. 2001. Large-scale analysis of the yeast proteome by multidimensional protein identification technology. *Nat. Biotechnol.* **19:** 242–247.

Wysocki V.H., Tsaprailis G., Smith L.L., and Breci L.A. 2000. Mobile and localized protons: A framework for understanding peptide dissociation. *J. Mass Spectrom.* **35:** 1399–1406.

Yang X., Dondeti V., Dezube R., Maynard D.M., Geer L.Y., Epstein J., Chen X., Markey S.P., and Kowalak J.A. 2004. DBParser: Web-based software for shotgun proteomic data analyses. *J. Proteome Res.* **3:** 1002–1008.

Zhang B., Chambers M.C., and Tabb D.L. 2007. Proteomic parsimony through bipartite graph analysis improves accuracy and transparency. *J. Proteome Res.* **6:** 3549–3557.

EXPERIMENT 9

High-throughput Cloning of ORFs
Assembling Large Sets of Expression Constructs

AN OVERVIEW OF FUNCTIONAL PROTEOMICS

In the past decade, biological research has witnessed a paradigm shift from focused reductionist approaches to an increasing reliance on genome-scale projects. These approaches capture data on a large scale and incorporate it into relational databases from which insight into biological systems, the organization of physiological networks, and new hypotheses can be derived. The explosion of genome sequencing projects is expanding our knowledge of genes and proteins and is dramatically increasing the number of potential candidate targets for the development of drugs, vaccines, and diagnostic markers.

It could be argued that the genome's main role is to harbor and maintain the information needed to construct cellular proteins. Among the most important outcomes from the genomic era will be its contribution to understanding the proteome and its many functions. Proteins constitute both the operating machinery and the bricks and mortar of cells. They are the key targets of therapeutics and the immune system and are frequently used as biomarkers for diagnostics. A fundamental challenge facing postgenomic biology will be the elucidation of protein function at the systems scale—a challenge that has led to the establishment of proteomics.

Even the simplest cell contains a large number of proteins, which makes studying the proteome of a cell a daunting task. The purview of proteomics encompasses many more unique species than any of its sister disciplines (e.g., genomics, transcriptomics, or metabolomics) by at least an order of magnitude. The burden of numbers is compounded by the complexity of protein behavior and the challenges inherent in studying them. Proteins are complex three-dimensional molecules, whose functions depend on the unique folding of each protein. Determining and measuring protein function requires a broad range of assays and approaches. Accomplishing this in a high-throughput format is the major challenge for the proteomic era.

Synergistic Approaches to Proteomics

The successful application of proteomics requires two complementary approaches, abundance-based proteomics and function-based proteomics. The predominant approach, abundance-based proteomics, typically utilizes mass spectrometry coupled with various protein separation techniques to examine specimens (tissues, body fluids, etc.) to identify the proteins present and, in particular, those with altered abundance under different conditions—for example, to identify proteins expressed highly in diseased tissue compared with normal tissue.

As discussed in previous chapters, the abundance-based approach faces some challenges. First, the methods of separation and detection strongly favor abundant proteins; more than one-half of the proteins in a sample may fall below common detection limits. Second, even when detected, there is no guarantee that a protein can be positively identified. Finally, once proteins with markedly different abundance are identified, it remains to establish their biological roles or to verify their potential value as therapeutic or immunological targets. Thus, it is clear that the abundance approach must be supplemented with information on protein *function*.

A more recent approach is functional proteomics, which is the focus of Experiments 9–11. This approach utilizes the high-throughput (HT) study of protein function by producing recombinant proteins and studying their characteristics, including structural features, protein–protein interactions, catalytic activities, drug interactions, biochemical activities, ability to produce an immune response, and as a source for producing antibodies and other useful reagents. In addition, the expression of proteins in vivo in screening experiments contributes to understanding the physiological consequences of protein expression, the suppression of genetic phenotypes, and the inherent properties of the proteins (e.g., subcellular localization, interacting partners, and posttranslational modification). Functional proteomics directly addresses the study of specific protein networks within cells or organisms and their functional pathways.

Functional proteomics avoids some of the limitations of the abundance approach. First, it is much less sensitive to the natural abundance of the protein. That is, even if a protein is ordinarily expressed in very low levels in vivo, its gene can be cloned readily and thus the protein can still be studied. Second, the identity of the protein is never in question. And, finally, because the focus is on protein function, experiments can specifically focus on finding proteins that display the activities of interest. However, the functional approach is limited both by potential artifacts introduced by the overexpression or ectopic expression of proteins and because it requires foreknowledge of the gene sequences and access to clones for the genes. Thus, by combining both approaches, the discovery features of the abundance approaches with the biological features of the functional approaches, an accelerated path to the discovery and annotation of all protein function can be achieved.

The Need for Flexible and Expression-ready Protein-coding Clone Collections

Nearly all methods for studying protein function begin with the expression of protein from cloned copies of the protein-coding sequences. Thus, obtaining a complete set of validated protein-coding clones is the first step toward establishing a functional proteomics platform for any organism of interest. A useful clone repository should have several features.

1. The collection should be indexed and addressable. The identity of the gene in each well or tube should be tracked by a database. This avoids wasting effort to determine ex post facto the identity of genes that are positive in a screening experiment and it allows for the collection, comparison, and sharing of data on each gene (even if there were no response).

2. The cloned collection should be "expression ready." HT functional assays often require the addition of peptide tags to enable purification (His6, GST), detection (GFP), or functional readout (Gal4-AD) of the test proteins. The addition of these peptides is only possible if the untranslated regions (UTRs) are removed and the open reading frames (ORFs) are all arranged in an identical reading frame.

3. The ORF clones should be captured in a flexible format that allows all possible functional assays. In practical terms, this means that it should be easy to transfer collections of protein-coding sequences simultaneously into any expression vector—in-frame and without mutation.

4. The collection should be fully sequence-verified and affordable.

FIGURE 1. Automated clone storage system. Large collections of ORF clones can be stored in and accessed easily from automated storage systems like this one. This system includes two automated –80°C freezers, each with 80,000-tube capacity, that are directly linked to the central picking station. The central picking station reads the individualized two-dimensional bar codes on the bottoms of the tubes and then uses a picking arm that selects and delivers the specific genes requested by the user.

5. The collection should be comprehensive. Ideally, every gene in the organism should be represented (Fig. 1).

Different protein functional studies demand different protein expression vectors, therefore a flexible vector system should be used that enables the cloned sequences to be transferred rapidly to any vector. This can be achieved most efficiently by using vectors that employ recombinational cloning, a strategy that allows DNA fragments flanked by site-specific recombination sites to be moved from one vector to another in a single-step procedure, in-frame and without mutation (Fig. 2). There is now extensive experience with several such systems including the Gateway system (Invitrogen) and the Creator system (Clontech). These reactions are simple enough to allow HT automation and are highly efficient. Once a "master" clone is created, the identical sequences can be transferred easily into all bacterial, mammalian, and viral vectors commonly used for protein analysis in vivo or in vitro.

The following protocols describe the process of transferring protein-coding sequences from a master vector into a protein expression (Destination) vector for use in functional experiments.

RECOMBINATIONAL CLONING

A destination vector is a vector modified to accept the protein-coding sequences in-frame from a master clone (Entry clone) using a recombinational cloning enzyme. Typically, it is a protein expression vector in which a cassette has been inserted in-frame at the position where the protein-

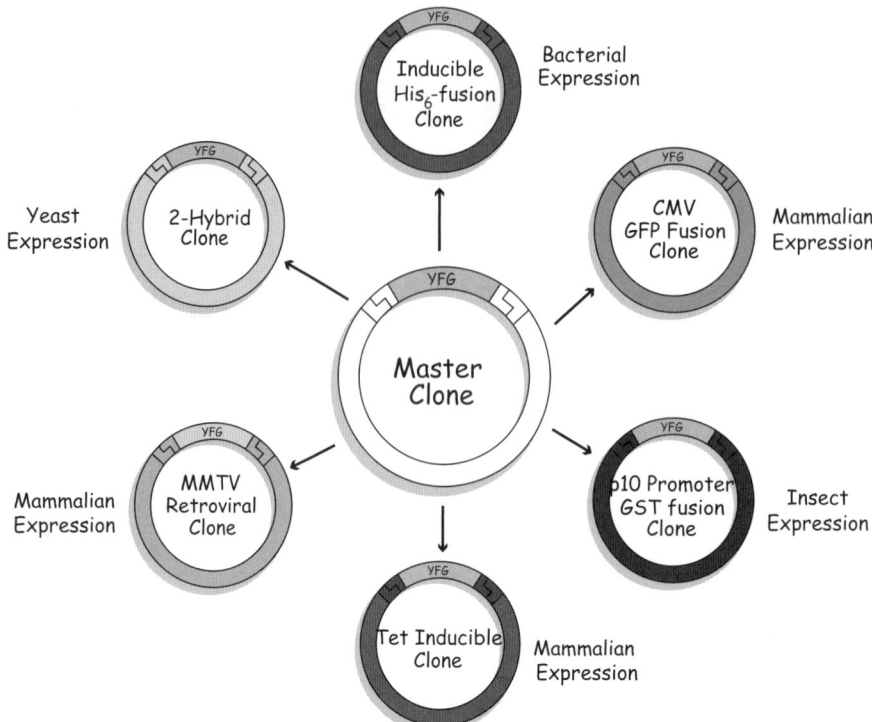

FIGURE 2. Master clone to expression clones. Once your favorite gene (YFG) is captured into a master vector to produce a master clone, a simple in vitro reaction can be used to transfer the insert into virtually any expression vector (that has the corresponding site-specific recombination sites) to create an expression clone. In this manner, proteins can be expressed in a wide variety of cell types, under the control of various promoters and with various tags added.

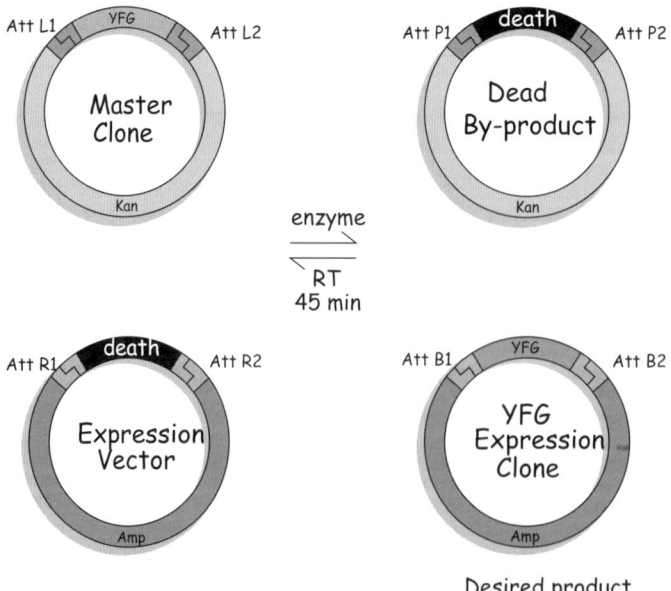

FIGURE 3. Site-specific recombination using the Gateway system. The master clone containing your favorite gene (YFG) flanked by site-specific recombination sites is mixed with an expression vector that carries a death gene flanked by the corresponding recombination sites. After adding enzyme and incubating, some fraction of the sites will recombine and resolve, creating a new product with YFG moved into the expression vector. In this example, the expression clone is the only plasmid that lacks the death gene and can grow on ampicillin.

coding sequences should be. The cassette comprises a counterselectable marker flanked by recombinational cloning sites that correspond to those in the master clone. When the master clone and destination vector are mixed with LR Clonase, the enzyme catalyzes a recombination event between the corresponding recombination sites in an orientation-specific manner (Fig. 3). By selecting for the marker on the destination vector and against the counterselectable marker in its cassette, only destination vectors that have swapped out their cassettes and replaced them with protein-coding sequences are obtained.

Ensuring that sequences remain in-frame enables destination vectors to add peptide tags to all inserted protein sequences. In the reaction below, the destination vector adds glutathione S-transferase (GST) to all of the expressed proteins, which enables the translated proteins to be captured to a solid surface (such as a bead or array surface) through an anti-GST antibody. This tag can work on either end of the protein, but if adding it to the carboxyl terminus is desired, it is important to use master clones in which the stop codon has been omitted from the protein-coding sequences or the ribosomes will terminate at the stop codon.

The protocols below use 96-well plates because they will be used later to build protein microarrays, which typically depend on printing hundreds or even thousands of genes. Our goal here is to describe how to operate these experiments efficiently in HT, where ease of operation and cost become considerable issues. If only a few samples need to be transferred, the recombination reactions can be executed using microfuge tubes. This is more efficient than the methods described here, but also too expensive for HT operations. Consult the user manuals that accompany the products for protocols that describe working with individual tubes.

PROTOCOL 1

Assembling Expression Constructs Using the Gateway LR Reaction

MATERIALS

Reagents

Destination vector
LR Clonase (Invitrogen)
LR Clonase buffer (Invitrogen)
Minipreps of Entry clones
(Optional) Topoisomerase I

> Topoisomerase I is expensive. It is equally efficient to linearize the destination vector using a unique restriction site within the recombination cassette between the attR sites.

Equipment

96-well polymerase chain reaction (PCR) plate containing master clones

> Each well will contain a different master clone. When the DNA for the master clone is prepared in HT there likely will be variation in concentration from well to well. Keeping this variation to a minimum is key to success in HT. Although the reaction conditions here are designed to be robust, they cannot compensate for a concentration range that is too broad.

PROCEDURE

1. Prepare the Master mix on ice. The amount below could be dispensed to 104 wells, and thus is sufficient for one 96-well plate.

Destination vector	104 μL
LR Clonase	104 μL
LR Clonase buffer	208 μL
Topoisomerase I	14 μL
Deionized H_2O	93.6 μL

   Prepare the destination vector (pANT7-GST, in this case) in large scale using a method that makes high-quality DNA, such as anion exchange. Avoid RNA and protein contamination, so that DNA quantification will be accurate. The final amount of DNA added for the destination vector is 12 ng per reaction.

2. Mix the contents of the Master mix thoroughly by pipetting it up and down. Aliquot 5 μL of Master mix per well of a 96-well PCR plate.

   This is best accomplished using a high-quality multichannel pipette (Fig. 4A). If many plates are to be processed, a liquid-handling robot offers better accuracy, better plate tracking, and a log file of all of the steps performed (Fig. 4B).

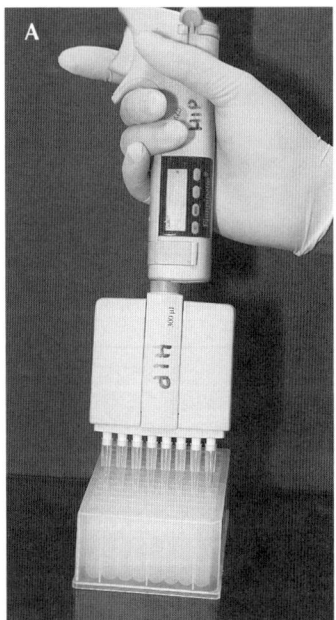

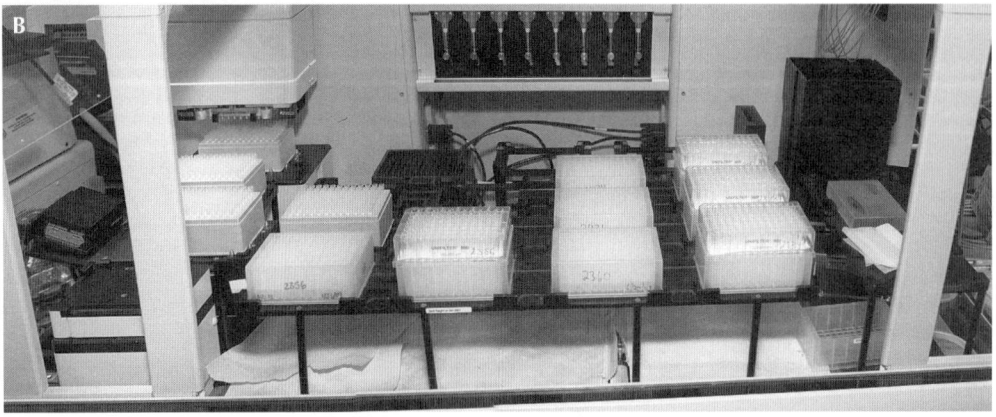

FIGURE 4. Tools for high-throughput liquid handling. (A) Multichannel pipettes like this are indispensable when doing high-throughput experiments. They can be used to resuspend many pellets or to add reagents to many wells. When adding the same reagent to many wells, a Master mix can be produced in a single tube and then divided up into one column of wells in a microtiter dish. This column can then serve as the source for filling all the wells of a 96-well dish. (B) If many plates need to be processed, liquid-handling robots are more accurate and faster and they never confuse wells or lose their places. Of course, they are also very expensive.

3. Add 5 μL (3 ng) of diluted Entry clone miniprep DNA to every reaction. These are called the donor clones.

   Miniprep DNA prepared using commercial kits or with in-house prepared reagents works well as long as the DNA is carefully quantified and the amount of DNA added is accurate. The DNA preparation method described below (isolation of DNA plasmids in 96-well plate format) works well for this step.

4. Incubate the reactions for 1–3 hours at 25°C.

   Although the user manual calls for room temperature, reactions incubated at a controlled temperature of 25°C are more robust and reliable.

5. Centrifuge the plate briefly to collect the reactions at the bottom of the wells.

6. Either transform the plasmids into competent bacteria (as described in the next protocol) or, if not proceeding immediately with transformation, freeze the reactions and store them at –20°C.

PROTOCOL 2

Transforming Chemically Competent Bacteria with the Gateway LR Reaction Product

MATERIALS

Reagents

Competent cells (for preparation, see Appendix 9)

5x KCM buffer (500 mM KCl <!>, 150 mM $CaCl_2$ <!>, and 250 mM $MgCl_2$ <!>)

LB agar plates (containing the appropriate antibiotic)

When working with 96-well plates, it is impractical to plate transformed bacteria onto standard 10-cm Petri dishes containing LB agar, because errors in tracking 96 individual dishes are guaranteed. Instead, our lab developed a 48-sector plastic insert that can be placed into a 25-cm x 25-cm square bioassay dish that holds 330 mL of melted LB agar. Thus, a single 96-well plate can be plated onto two 48-sector dishes, a configuration that is compatible with plating and picking robots (Fig. 5). The dish and insert are available from Genetix (part X6027).

LR reaction products (these plasmids are from Step 6 of the previous protocol)

2 M Magnesium solution (use to prepare SOB medium)

Prepare 100 mL.

Component	Amount to Add	Final Concentration
$MgSO_4 \cdot 7H_2O$ (Fisher M-63)	24.6 g	1 M
$MgCl_2 \cdot 6H_2O$ (Sigma M-9272)	20.3 g	1 M

SOB medium (use to prepare SOC medium)

Prepare 1 liter of media in a 2-liter flask.

Component	Amount to Add	Final Concentration
yeast extract	5 g	0.5%
Tryptone	20 g	2%
NaCl	0.58 g	10 mM
KCl	0.18 g	2.5 mM

Dissolve ingredients in MilliQ-H_2O. Autoclave for 15 minutes at 250°F (121°C) and 15 psi. Add 10 mL of a 2 M magnesium solution (final concentration = 20 mM).

SOC medium

Add 10 mL of 2 M D-glucose to 1 liter of SOB medium. Sterilize the solution by filtering through a 0.22 μm filter.

FIGURE 5. Plating bacteria on bioassay dishes. Bioassay dishes with inserts like these enable the plating of 48 different clones onto a single plate instead of 48 separate Petri dishes. As shown in the *inset*, these plates are also compatible with automatic picking robots.

Equipment

96-well PCR plate
Gas-permeable seal for plate
Incubator set at 37°C

PROCEDURE

1. When all of the necessary reagents are assembled and prepared, thaw the competent bacterial cells in an ice slurry.
2. Prepare Transformation Master mix by combining 16 µL of autoclaved distilled, deionized H_2O and 4 µL of 5x KCM buffer per reaction.

For each 96-well plate, prepare enough Transformation Master mix for 100 reactions, thus allowing a little margin for error. Thus, 1.6 mL of H_2O plus 400 µL of 5x KCM buffer total mix would work well.

3. Aliquot 20 µL per well in a PCR plate and incubate the plate on an ice slurry for at least 5 minutes.

   If this is done manually, it is easier to use an electronic pipette that can aspirate enough mix to aliquot to multiple wells (Fig. 6).

4. Keeping the plate on ice, transfer 25 µL of thawed competent cells to each well of the PCR plate containing Transformation Master mix.

   If this is done manually, it is easier to use an electronic pipette that can aspirate enough mix to aliquot to multiple wells. Remember to prepare stocks with an excess of at least 5 µL because of the challenge in pipetting fluids with high levels of glycerol.

   It saves time to have the competent bacterial cells aliquoted and frozen in the 96-well format.

5. Add 5 µL of LR reaction product. Mix well by pipetting up and down.

   This step can be done using a multichannel pipette. The LR mix is added last because it is unique for each well. The pipette tips must be changed after each addition.

6. Incubate the plate for 20 minutes on ice.

7. Remove the plate from the ice and let it stand on the benchtop for 10 minutes at room temperature.

8. Return the plate to the ice and incubate it for 2 minutes.

9. Add 100 µL of SOC medium. Seal the plate with a gas-permeable seal and incubate it for 60 minutes at 37°C.

   Addition of SOC medium can be done using a multichannel pipette, being careful not to let the tip touch any of the wells so as to avoid cross contamination. Shaking the plate during the incubation is optional.

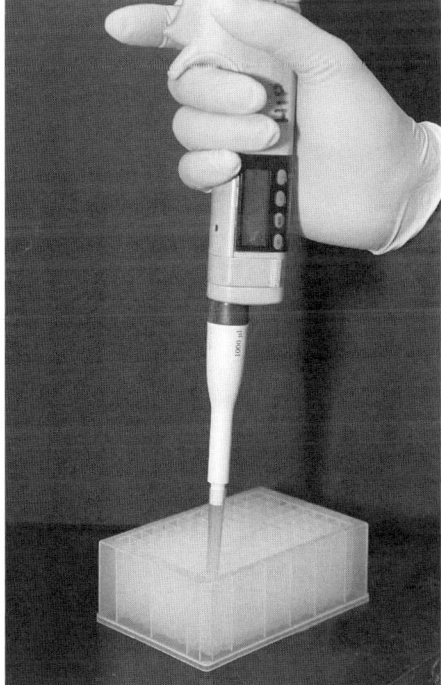

FIGURE 6. Electronic pipettes enable the rapid addition of samples to many wells. When adding the same thing to many wells, electronic pipettes like this are very handy. A large volume can be aspirated into the tip and then dispensed into each well without going back to the master tube until the end of a row or column.

10. Plate 100 μL of each reaction onto LB agar plates containing antibiotic for selection.

    The Transformation mix should be added to each sector of the 48-sector plate. After each column of eight sectors has been inoculated, lift the plate and rock it in a circular motion to spread the mix on the sectors. A spreading device is not needed.

    Important: Change pipette tips between samples.

    This step can be completed more quickly if it is done with a liquid-handling robot that can allow variable spacing between liquid dispensing tips.

11. Incubate the plates for 20 hours at 37°C.

    We typically begin the incubation with the plate lid face up for the first hour or two. The plate is then inverted for the remainder of the time, to prevent condensate from dripping onto the agar.

EXPERIMENT 10

Construction of Protein Microarrays
Nucleic Acid Programmable Protein Array (NAPPA)

Functional proteomics enables protein activities to be studied in vitro using high-throughput (HT) methods. Protein microarrays are the method of choice because they display many proteins simultaneously and require only small reaction volumes to assess function. Protein microarrays are typically used to (1) measure the abundance of many different analytes in a sample or (2) study the functions or properties of many proteins spotted on the array.

Abundance-based Microarrays

Abundance-based microarrays come in two forms: capture arrays and reverse-phase protein blots (Fig. 1).

Capture arrays comprise an array of analyte-specific reagents (ASRs)—almost always antibodies—that have known specificities and can capture their cognate antigens from a complex sample for use in measuring abundance. There are two basic approaches for measuring analyte abundance. The first is direct labeling, in which all the analytes in a sample are labeled (e.g., radioactively or with a fluorescent tag). This allows simultaneous measurement of many different analytes but demands exquisite specificity of the ASRs on the array. This scheme also suffers if the labeling

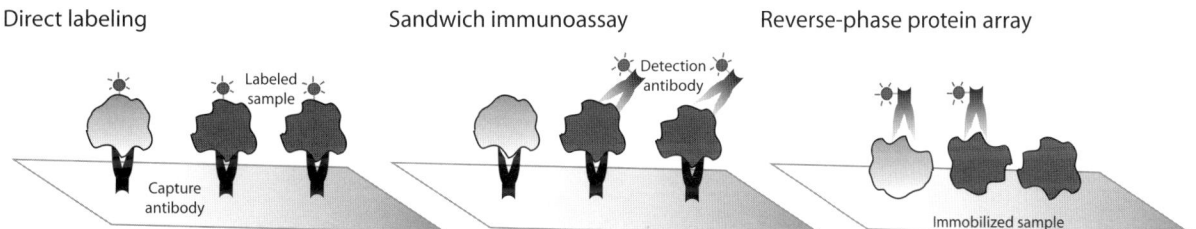

FIGURE 1. Detection methods for abundance-based protein microarrays. Capture or antibody microarrays can be analyzed by chemically modifying the sample with readout markers before applying the sample to the microarray (*left*). This has the advantage that all features of the microarray can be assessed simultaneously, but any cross-reactivity of the antibodies used to capture the analytes will give false readings (*light gray feature*). Alternatively, the analytes can be detected by capture with one antibody and detection with a second specific to a different epitope in sandwich immunoassay fashion (*middle*). This significantly reduces false readouts (*light gray feature*), but can be more cumbersome to multiplex. In the reverse-phase protein blot (*right*), the complex experimental sample itself is printed and probed with an antibody. This allows the rapid screening of many samples, but is highly subject to the characteristics of the detecting antibody.

method is not uniform. The second method uses an additional ASR, such as a fluorescently tagged antibody, that binds to a different epitope on the captured antigen creating a sandwich assay that provides excellent specificity. However, this approach significantly limits the number of analytes examined simultaneously and requires at least two good ASRs to different epitopes on each measured analyte.

Unlike other arrays, *reverse-phase protein blots* (RPPBs) are not arrangements of known elements with defined content but rather comprise printed spots of the unknown samples themselves. In some cases the samples are printed directly, although for very complex samples the limited number of binding sites in the surface area of the feature may lead to an overrepresentation of abundant species. Thus, increasingly the samples are first prefractionated using a variety of biochemical methods, and the various fractions are then printed. Prefractionation reduces the complexity of each printed spot, thus increasing the chances of observing less abundant species. Signals are detected in RPPBs by probing the blotted samples with specific ASRs. Thus, like the capture arrays, this technique depends heavily on having good-quality, highly specific ASRs for the analytes of interest.

Abundance-based protein microarrays are the most common form of protein microarray and were the first to be commercially exploited because there were moderate-sized collections of commercial antibodies available that can be spotted on the arrays. This approach has great potential as a profiling and screening tool. The major challenge here is the lack of a broad range of antibodies that can be used. The majority of these arrays are cytokine arrays because there are such good collections of cytokine antibodies available. To be useful at the proteome scale, antibodies to every human protein (including their isoforms) must be produced. Currently, only a small fraction of such antibodies are available and even fewer antibodies have the appropriate characteristics to function in the context of a protein microarray. Another limitation of the abundance-based capture arrays is that they do not provide any information about protein function.

Function-based Arrays

This section will focus on *function-based arrays* or *target protein arrays*, which are microarrays on which are printed the arrayed proteins of interest. Each feature on the array has a known protein printed on it, called a target protein, whose address is documented on an array map. These target protein arrays can be probed with a large variety of *query* molecules in order to examine protein interactions with other proteins, drugs, nucleic acids, lipids, or antibodies. Moreover, target arrays can be used to examine enzyme–substrate relationships, for example, to examine which proteins on the array are substrates of an active kinase. The list of potential applications of such microarrays is large. A partial list includes:

1. Building protein interaction networks, including the assembly of multiprotein complexes, to shed light on biochemical pathways and networks.
2. Screening proteins expressed by pathogenic organisms with serum from convalescent patients to identify immunodominant antigens to predict good vaccine candidates.
3. Testing potential interactions of a drug with a broad range of proteins to look for unintended binding targets that might suggest possible toxicities.
4. Testing the selectivity of drug binding to a target class of related proteins.
5. Mapping the substrate profile for active enzymes against a broad range of protein targets.
6. Testing a mutant series of a particular protein in a single functional experiment to map critical amino acids or functional domains.

7. Testing the serum from patients with autoimmune or chronic diseases against candidate antigens to identify autoantibodies and biomarkers.

First-generation studies on building and employing protein microarrays have shown significant promise with respect to useful applications of the technology. MacBeath and Schreiber (2000) immobilized purified proteins on aldehyde-treated glass slides to examine protein interactions between several known interacting protein pairs, an approach later exploited to examine protein interactions among the basic-region leucine zipper (bZIP) transcription factors (Newman and Keating 2003). Zhu et al. (2001) expressed, purified, and immobilized a 6x histidine-glutathione-S-transferase (GST) tagged version of the *Saccharomyces cerevisiae* proteome on nickel-coated glass slides and used the arrays to screen for calmodulin-binding proteins and phosphoinositide-binding proteins. Additional studies have demonstrated that protein microarrays can be used to detect enzyme substrates and to find the targets of immune responses (Ptacek et al. 2005; Zhu et al. 2006). Thus, protein–protein, protein–drug, protein–antibody, protein activity, and protein specificity have all been successfully evaluated using this format, demonstrating an incredibly broad range of the applications available for this powerful tool.

Among the greatest challenges for target protein microarrays is their manufacture. Currently, target protein microarrays are generated by expressing, purifying, and spotting the proteins onto a solid surface at very close spatial density. To produce arrays with hundreds or thousands of proteins necessitates access to HT methods of protein production. Unlike DNA, there are no simple enzymatic methods for rapidly copying and amplifying proteins. Instead, the expression and purification of proteins are tedious, costly, and frequently unsuccessful for many target proteins. The use of HT methods lowers the sample volumes and thus the overall yield of the final product. Moreover, production of protein microarrays must be accomplished using procedures that preserve the activity of the proteins on the array. Proteins are notoriously fragile and are often damaged during the many manipulations required to produce protein microarrays. Once printed on the arrays, the proteins tend to unfold and lose their activity. It is impossible to predict which proteins will lose activity, and thus the shelf life of the arrays, even kept frozen, may be only a matter of days.

An alternative approach is to translate the proteins in situ on the array surface. This approach, termed Nucleic Acid Protein Programmable Array (NAPPA), enables the simultaneous expression of proteins in microarray format without the need for individual protein purification. This method uses cell-free extracts (typically T7 polymerase and reticulocyte lysate) that transcribe and translate DNA into proteins, thus converting cDNA (copy DNA) copies of genes into the desired target proteins (Fig. 2). Instead of printing proteins at each feature of the array, the cDNA molecules for the corresponding genes that produce desired proteins are affixed to the array. The cDNAs have all been configured to include an epitope tag at one end of the protein. Along with the DNA, a capture agent, such as an antibody, is also printed so that when the protein is produced, it is immediately captured to the array's surface. Once the various genes and the capture agent are printed, the arrays are stable and can be stored dry.

To activate these self-assembling protein microarrays, the in vitro transcription/translation (IVTT) mix is added, the various proteins are then transcribed and translated by the molecular machinery in the extract, and when the proteins are produced, they are captured to the array surface. At this point, the extract can be washed away and the array can be employed in experimentation. This approach eliminates the need to express and purify proteins separately and produces proteins "just in time" for the assay, abrogating concerns about protein array shelf life. This chemistry also has the advantage that mammalian proteins can be expressed in a mammalian milieu (reticulocyte lysate) to increase the efficiency of expression and to encourage natural folding of the proteins, providing access to vast collections of cloned cDNAs.

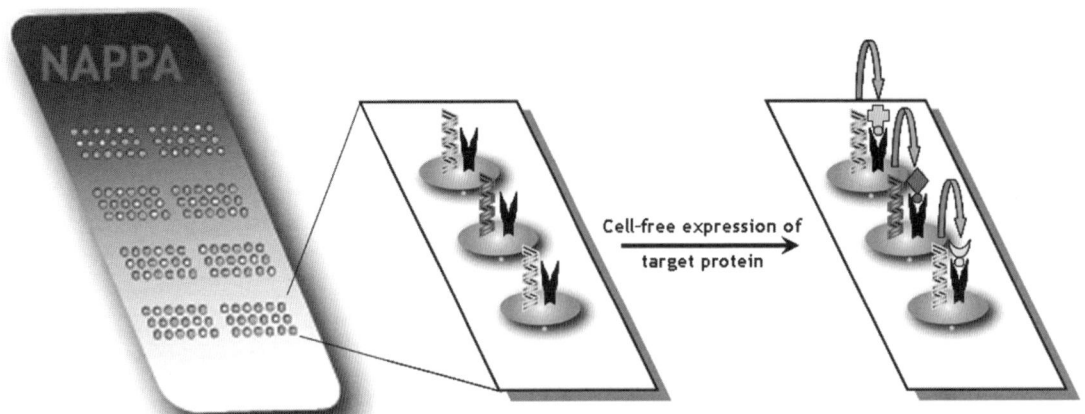

FIGURE 2. Schema for Nucleic Acid Programmable Protein Array (NAPPA). Instead of purifying proteins and then printing them on the arrays, NAPPA entails printing DNA encoding the protein and then transcribing and translating the proteins in situ on the array surface. By this method, the central dogma of molecular biology is recapitulated at each array feature.

The recombinational cloning approach described in Experiment 9 is an ideal method for creating the cDNA content for NAPPA. All of the genes in a master clone collection can be transferred into a vector designed for in vitro transcription and translation and which adds an appropriate epitope tag.

Basics of the NAPPA Chemistry

The methods described here for building self-assembling protein microarrays were built around several considerations regarding the features that will be important for experimental applications. First, it was desirable to establish a high-density format that minimized the use of cell-free extract and allowed the simultaneous examination of many proteins. This will eventually reduce the cost per protein. Second, the method was designed to use a readily available matrix (such as standard glass microscope slides) that did not require micromachined nanowells. This both reduces the need for special instrumentation needed to interface with the matrix and makes the method more generally available. Third, the method was designed to utilize the widely accessible existing technologies for printing and reading DNA microarrays. Generally, any solid pin microarraying device should suffice for printing and most fluorescent readers for DNA microarrays will also work for reading the protein arrays. This avoids the need to create specialized equipment to produce and print the arrays and will hopefully ensure broad accessibility of the technology. Fourth, it was important that sufficient protein was captured at each feature to enable the study of protein function. Using these methods, the average feature on NAPPA has about 50 fmole of protein. Finally, the method required an efficient printing chemistry that supported transcription and translation in situ. Once the proteins are translated, this chemistry had to display rapid, efficient, and specific protein capture, without high background signal and without spot-to-spot diffusion or cross talk.

As most arrays will entail hundreds to thousands of different DNA species, the printing chemistry must be amenable to HT processing of the target plasmid DNA and must attain sufficient DNA binding without sacrificing the integrity of the plasmid. Early experiments demonstrated that supercoiled plasmid performed much better than nicked or linearized template at transcription/translation. A variety of DNA derivatization schemes were originally tested (direct UV absorption, incorporation of surface reactive nucleotides, cross-linking agents, DNA-binding proteins, etc.); some captured DNA poorly, whereas others captured DNA well but yielded poor protein expression.

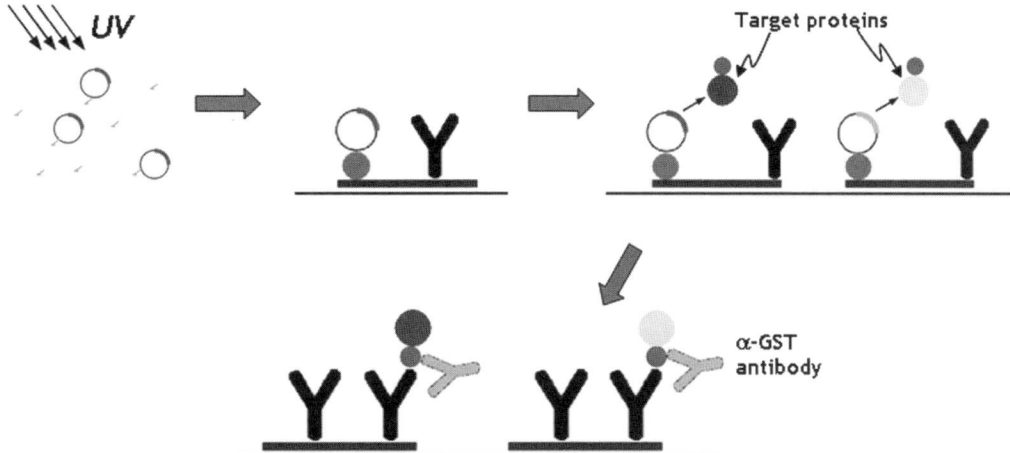

FIGURE 3. Schema for chemistry used to print NAPPAs. Plasmid DNA encoding different proteins with GST fusions is crosslinked to a psoralen–biotin conjugate using UV light. Avidin, polyclonal GST antibody, and linker are added to the biotinylated plasmid DNA and the samples are arrayed onto glass slides. Microarrays are then incubated with rabbit reticulocyte lysate with T7 polymerase to express the proteins that all have GST tags. Target proteins are expressed and immobilized by the polyclonal GST antibody. All target proteins are detected using a monoclonal anti-GST antibody against the carboxy-terminal tag confirming expression of full-length protein.

The most efficient strategy that emerged was to couple a psoralen–biotin conjugate to the expression plasmid DNA using UV light and then capture this on the surface through avidin (Fig. 3).

The protein is captured through an epitope tag. In most instances, the addition of a carboxy-terminal GST tag to each protein through recombinational cloning is used to capture it to the array through an anti-GST antibody printed simultaneously with the expression plasmid in a 15-fold molar excess. Other protein fusion tags and capture molecules can be substituted easily for the GST fusion and anti-GST antibodies. The advantage of a carboxy-terminal tag is that full-length protein can be confirmed by using a separate antibody to the GST tag.

Once the array is printed, it can be dried and stored at room temperature without significant loss of signal for months. To activate and use the array, a cell-free, coupled transcription/translation system (such as reticulocyte lysate containing T7 polymerase) is simply added as a single continuous layer covering the arrayed cDNAs on the microscope slide. This avoids the need to deliver the expression system to each individual position on the array.

This chemistry combines the advantage of a high-affinity biotin and avidin/streptavidin interaction with the ease of incorporating psoralen into double-stranded DNA. Fortunately, biotinylation and immobilization of the plasmid DNA is efficient over a wide range of pH and salt concentrations. An estimate for protein yield can be made by comparing the anti-GST signal for the expressed and captured protein to a signal obtained from printing a known amount of purified GST protein. From this, NAPPA captures about 400–2700 pg/feature or about 50 fmole/feature on average.

Efficiency of protein expression and capture requires a balance between efficient capture of DNA to each feature (which increases with high psoralen–linker concentration) and efficient transcription/translation (which is inhibited as the number of intercalating psoralen molecules increases per plasmid). The linkers described below were the best of a variety of tested psoralen–biotin linkers with chemical configurations that are less likely to affect transcription/translation. The conditions described represent a good balance between efficient DNA capture and accessibility for T7 transcription. Fortunately, there is a relatively broad range making the following conditions fairly robust (Fig. 4).

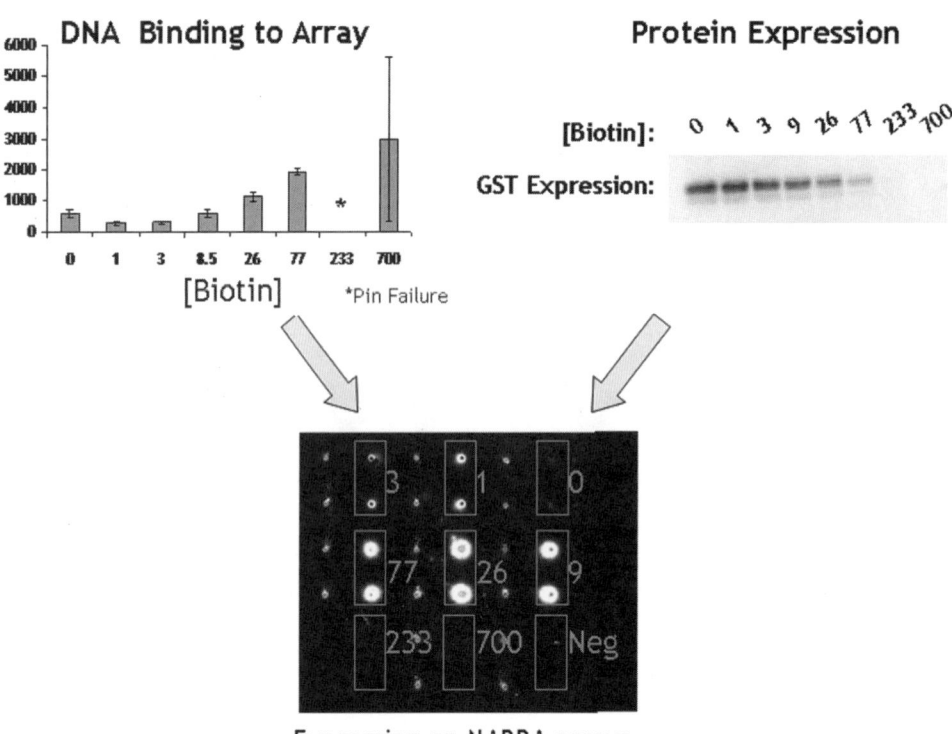

FIGURE 4. Balance of conditions affects protein yield on the microarrays. Increasing biotin concentrations leads to better DNA capture on the arrays but inhibits protein expression at high levels (presumably via inhibited transcription from the plasmid). Protein levels on the arrays can be optimized by balancing the two.

Producing NAPPA Protein Microarrays

The protocols that follow provide details on preparing, expressing, and detecting NAPPA microarrays. The glass slides are prepared and bacterial transformants inoculated on Day 1. Plasmids are isolated the following day and arrayed to make NAPPA slides. On Day 3, NAPPA slides are expressed, and on the last day, proteins and DNA are detected using antibodies and stains.

PROTOCOL 1

Coating Glass Slides with Amino Silane

MATERIALS

CAUTION: See Appendix 11 for appropriate handling of materials marked with <!>.

Reagents

2% Amino silane (3-aminopropyltriethoxysilane) <!> in acetone <!>
   Prepare 300 mL of 2% amino silane by adding 6 mL of amino silane (Pierce, 80370) to 294 mL of acetone.

Equipment

Glass box
Glass slides (VWR, 48311-702)
Lock & Lock 1.5-cup boxes (Heritage Mint, ZHPL810)
Metal 30-slide rack (Wheaton Science Products, 900234; handle removed)
Rocking shaker

PROCEDURE

1. Place 30 slides into the metal slide rack. Treat the glass slides in 2% amino silane for approximately 5 minutes (the range is 1–15 min) inside a glass box on a rocking shaker.

2. Rinse the slides with acetone while they remain in the rack.

3. Briefly rinse them with deionized H_2O. Dry them with filtered compressed air.

4. Store the silanized slides at room temperature in their rack in a storage box.

PROTOCOL 2

Preparing Bacterial Cultures in a 96-Well Format

MATERIALS

Reagents

Bacteria containing an expression plasmid
   Bacteria are either on agar plates or in 96-well format on agar or in glycerol medium.
KPi (potassium phosphate)
   Prepare KPi by combining 23.1 g of KH_2PO_4 (FW 136.09) and 125.4 g of K_2HPO_4 (FW 174.18) in deionized H_2O. Adjust the volume to 1 liter with deionized H_2O. Filter sterilize the solution. (Final concentrations are 0.17 M KH_2PO_4 and 0.72 M K_2HPO_4.)
Terrific Broth
   Prepare 900 mL of TB in a 2-liter flask.

Component	Amount to Add	Final Concentration (per liter)
Yeast extract	24 g	2.4%
Bacto-tryptone	12 g	1.2%
glycerol	4 mL	0.4%

   Dissolve the ingredients in 900 mL of H_2O. Autoclave for 15 minutes at 250°F (121°C) and 15 psi.
Terrific Broth (TB) culture medium
   Combine 900 mL of TB and 100 mL of KPi. Just before using (in Step 1), add 10% ampicillin to a final concentration of 100 ng/mL.
   Ampicillin is mentioned here because nearly all NAPPA vectors use this as a selectable marker. The appropriate drug should be substituted as needed for other vectors. Failure to express and capture protein on microarrays can usually be attributed to poor DNA yield, so it is important to use plasmids that give high yields.

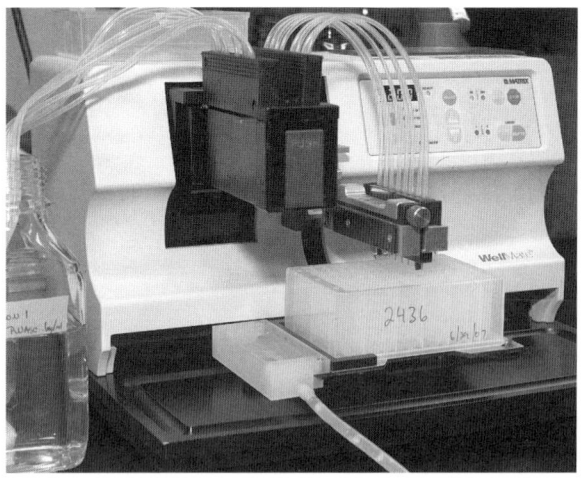

FIGURE 5. Rapid well-filling devices. Instruments like this one use peristaltic pumps to rapidly fill wells in 96-well plates with liquid from the reservoir on the left. They are much less expensive than liquid-handling robots (see Experiment 9, Fig. 4B) and are compatible with stackers so that many plates can be processed without having to stand at the machine. In addition, the tubes can be separated so that each row of the plate can get a different solution.

Equipment

96-pin transfer device

96-well deep-well block (Marsh/ABgene, AB-0661)

Centrifuge (with swinging bucket rotor that can hold a deep-well block plus the filter plate, adjustable speed, maximum 3750 rpm)

Gas-permeable plate seal (Marsh/ABgene, AB-0718)

LB agar plate (prepare in an OmniTray, Nalge Nunc, 62409 600)

Liquid pipetting device (e.g., Multidrop [Thermo Scientific] or Matrix WellMate [Thermo Scientific])

> The Thermo Multidrop and the Matrix WellMate are examples of a family of instruments that provide semiautomated liquid dispensing without having to purchase very expensive robots (Fig. 5). Typically, these devices have eight channels, corresponding to the eight rows on a 96-well microtiter plate, and they use a peristaltic pump to bring fluid from a reservoir and dispense it into the wells of the plate. They can deliver volumes in the range of 10 µL to 1 mL and can often fill a plate in a matter of seconds. The intake tubing can be rinsed and transferred to a different reservoir to change buffers.

Shaker for deep-well blocks set at 37°C

Vacuum manifold for 96-well plates (Eppendorf)

PROCEDURE

1. Add 1.5 mL of TB culture medium to each well of a deep-well block.

   If available, use an automated system (e.g., Matrix WellMate) to dispense the medium to the wells.

2. Inoculate each well with bacteria containing an expression plasmid.

   Inoculating from a glycerol stock in 96-well plates

   a. Sterilize a 96-pin device in 80% ethanol and then flame it. Let the pins cool.

   b. Using the 96-pin device, transfer bacteria from glycerol stocks to an LB agar plate (prepared in an OmniTray) that contains 100 ng/mL ampicillin. Allow the bacteria to grow overnight at 37°C.

c. Using the 96-pin device, inoculate the liquid cultures from the patches of bacteria. Ensure that individual bacterial colonies are visible (Fig. 6), indicating that the bacteria are not overgrown.

> To obtain maximum yields, we find that it is better to force the bacteria to grow overnight on agar under selection before inoculating the liquid medium. Although this adds an extra day, this routinely gives higher yields than going directly from glycerol stock to liquid medium, presumably because growth on solid medium forces bacteria to maintain high copy numbers of the plasmid. Alternatively, inoculate the culture block directly from the glycerol plate.

Inoculating from bacteria on agar

a. Using either toothpicks or a 96-pin device (depending on the growth configuration of the bacteria), transfer bacteria from the agar to the liquid medium in the deep-well block.

3. Cover the deep-well block with a gas-permeable seal.

4. Incubate the cultures on a shaker for 22–24 hours at 37°C, at either 300 rpm (large orbit) or 900 rpm (small orbit), depending on the shaker.

5. Pellet the bacteria by centrifugation for 15 minutes at 2000–2500 rpm.

6. If the bacteria will not be used immediately, then freeze the pellets at –20°C. Otherwise, proceed to the DNA preps (see Protocol 3).

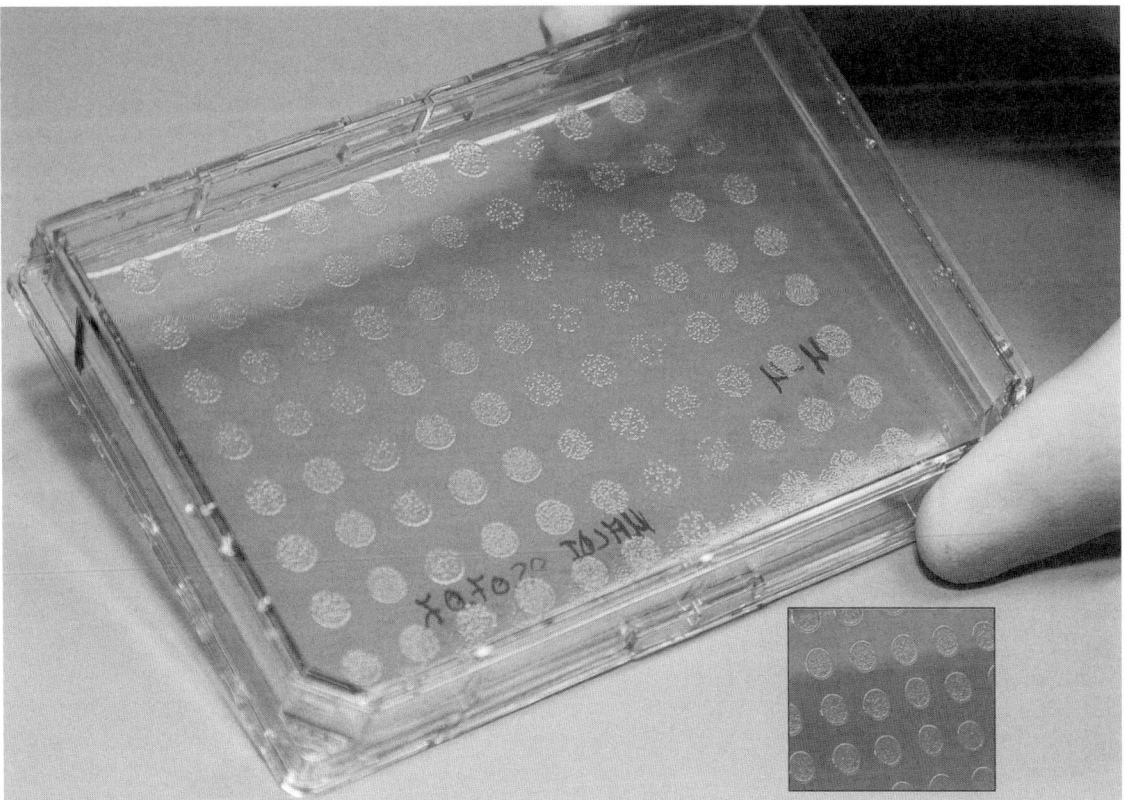

FIGURE 6. OmniTray. The OmniTray is a plate that has the same shape as a 96-well microtiter dish but has no wells. This is a convenient plate for creating a grid of bacteria that can be used with a 96-pin transfer device to inoculate 96-well cultures. Make sure that colonies are visible as shown. If the cultures are overgrown as shown in the *inset*, the yields will suffer.

PROTOCOL 3

Isolating DNA Plasmids in a 96-Well Plate Format

MATERIALS

CAUTION: See Appendix 11 for handling of materials marked with <!>.

Reagents

Bacterial pellet grown in a 96-well block (from Step 6 of Protocol 2)
Isopropanol (600 μL/well)
Solution 1: Resuspension buffer (200 μL/well)
    50 mM Tris (pH 8.0)
    10 mM EDTA (pH 8.0)
    0.1 mg/mL RNase
    Store at 4°C for up to 1 week.
Solution 2: Lysis solution (NaOH/SDS) (200 μL/well)
    0.2 N NaOH <!>
    1% (w/v) SDS (sodium dodecyl sulfate) <!>
    Prepare fresh each time.
Solution 3: Neutralization solution (200 μL/well)
    3 M potassium acetate (pH 5.5)
    Dissolve 294.4 g of potassium acetate in 800 mL of deionized H_2O. Add glacial acetic acid <!> until the pH is 5.5 (~115 mL). Adjust the volume to 1 liter with deionized H_2O, and store the solution at 4°C.

Equipment

96-well deep-well block (Marsh/ABgene, AB-0661)
Aluminum plate seals (need five per 96-well block; Beckman, 538619)
Glass fiber 25-μm filter plate (Whatman, 7700-2804)
Liquid pipetting device (e.g., Multidrop [Thermo Scientific] or Matrix WellMate [Thermo Scientific])

PROCEDURE

1. Add 200 μL of Solution 1 to each pellet of bacteria and resuspend them by vortexing or pipetting up and down (Fig. 7).

    Solution 1 is probably best added with a single- or a multichannel pipette (for subsequent steps, the Matrix WellMate can be used). Whatever is used, mixing the sample is essential. Aside from strong antibiotic selection, adequate resuspension of the bacteria is the most important determinant for achieving a good yield of DNA. If bacteria remain at the bottom of the well or remain partially clumped, the yields will drop significantly. As the yields drop, the relative level of other contaminants rises. After resuspension, the sample should look silky. Bacteria are stable in Solution 1, so there is no hurry to get to the next step.

2. Add 200 μL of Solution 2 to each well, seal the plate with an aluminum plate seal, and gently invert the plate five times.

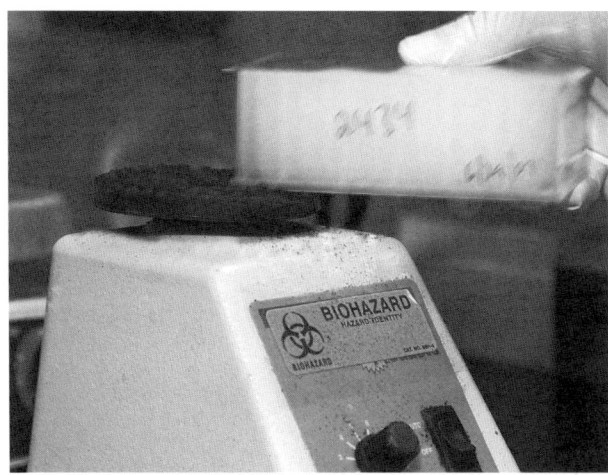

FIGURE 7. Vortexing pellets to resuspend bacteria. The best method for resuspending the bacteria is to pipette up and down in each well. If vortexing is used it must be done thoroughly until the pellets can no longer be seen.

After mixing, the samples should turn clear and viscous (Fig. 8). Work with haste, because waiting too long, particularly at warmer temperatures, can lead to chromosomal DNA breakage and consequent contamination of the sample. The entire step should take no longer than 5 minutes at room temperature.

3. Add 200 µL of Solution 3 to each well, seal the plate with an aluminum plate seal, and invert the plate ten times.

   The samples will take on a watery cottage cheese appearance. There is no need to wait at this point; centrifuge the samples immediately.

4. Centrifuge the block for 10 minutes at approximately 4000 rpm.

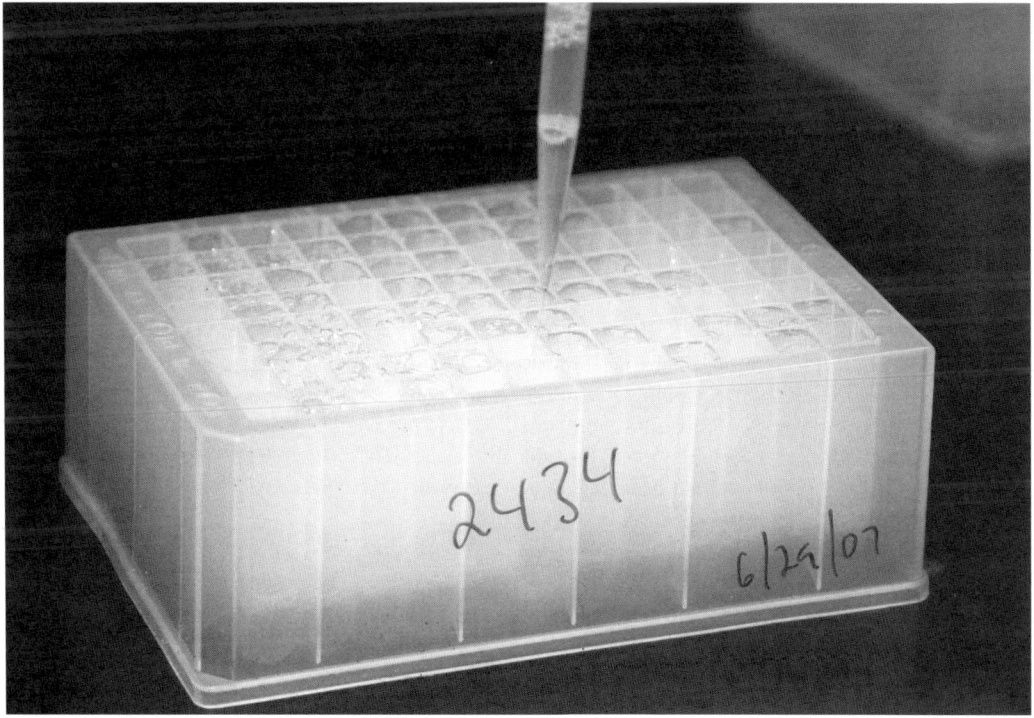

FIGURE 8. Clarified lysate. After the addition of the NaOH/SDS solution, the samples should become clear and viscous.

FIGURE 9. Stacked filter plate on top of deep-well block with isopropanol.

5. While the DNA preps are centrifuging, add 600 μL of isopropanol to each well of a deep-well block. (This can be done with the Multidrop or Matrix WellMate.) Stack the 25-μm filter plate onto the top of the deep-well block containing the isopropanol (Fig. 9). Set aside, but keep at room temperature.

6. Using a multichannel pipette, transfer the lysate supernatant from the centrifuged block (Step 4) to the filter plate. Be very careful to maintain the mapping (make sure the A1 solution from the source plate goes to the A1 well in the destination plate). The pellets are not packed tightly, so work quickly and avoid the cell debris.

7. Spin the stacked plates at 750g for 10 minutes at 25°C to filter out the remaining cell debris and transfer the lysate to the isopropanol.

8. Remove the 25-μm filter plate. Seal the block with an aluminum plate seal and invert the block several times to mix.

9. Centrifuge the block at maximum speed for 20 minutes at 25°C.

    Isopropanol will precipitate salt at cold temperatures, so this spin must be at 25°C.

10. Discard the supernatant by dumping the block over a waste container and tapping it several times onto dry paper towels.

11. Invert the block onto the bench for 10 minutes to allow the samples to dry. Do not allow them to dry for too long, or the DNA pellet will not dissolve in the next step (Fig. 10).

    It is important to remove all of the isopropanol prior to dissolving the pellet because isopropanol will contaminate the DNA prep. Drying time may vary according to the environmental humidity. A good indicator is the isopropanol smell. Redissolve the DNA pellet as soon as you cannot smell any isopropanol in the wells. This typically takes 5–10 minutes. Keep the block upside down for drying but propped up at a slight angle to allow air to circulate inside the wells.

12. Dissolve each DNA pellet in 150 μL of deionized H_2O. Shake the block for at least 15 minutes at room temperature.

13. Cover the block with an aluminum plate seal. Centrifuge the block briefly to collect the samples at the bottom of the wells.

14. Quantify the amount of DNA in each well using the procedure in Appendix 10 (DNA Quantification).

FIGURE 10. Inverted deep-well block to dry isopropanol-precipitated DNA pellets.

Adequate amounts of DNA are needed for successful protein expression and capture. A minimum of 30 µg of total DNA in the 150 µL reaction (200 ng/µL) is required. If the yields are insufficient, repeat the DNA production before moving to the next step. Two separate minipreps could be prepared and combined, but this might increase the amount of contaminants. Ideally, prepare sufficient DNA from a single prep.

15. Proceed to Protocol 4. If the samples will not be used immediately, freeze the block at –20°C.

PROTOCOL 4

DNA Biotinylation, Precipitation, and Arraying of Samples

MATERIALS

CAUTION: See Appendix 11 for appropriate handling of materials marked with <!>.

Reagents

BS3 linker (bis[sulfosuccinimidyl]suberate; Pierce, 21580)
DNA samples (from Protocol 3)
Ethanol (80%, room temperature)
GST protein, purified (0.6 mg/mL)
Isopropanol (room temperature)
Polyclonal anti-GST antibody (Amersham, 27457701V)
Psoralen–biotin <!> (Pierce, 29986)
Sodium acetate (3 M, pH 5.5)
Streptavidin
Whole mouse IgG antibody

166 EXPERIMENT 10

Equipment

96-well plate (Greiner, 651201)
>This is a high-volume, conical bottom plate used to augment recovery of the DNA pellet.

384-well plate for arraying (Genetix, X7020)
>See the note at Step 15.

Centrifuge (e.g., Eppendorf, 5810)
Humidified chamber (see Step 20)
Lock & Lock 1.5-cup boxes (Heritage Mint, ZHPL810)
Metal 30-slide rack (Wheaton, 900234, handle removed)
Orbital shaker
Solid pin arraying device (e.g., QArray Mini)
UV cross-linking device (e.g., Stratagene 1800 Crosslinker)

PROCEDURE

1. Transfer the entire DNA solution (150 µL per well) from each well of the deep-well block to the Greiner 96-well plate.

    As mentioned above, adequate amounts of DNA are needed for successful protein expression and capture. A minimum of 30 µg of total DNA in the 150 µL reaction (200 ng/µL) is required.

2. Prepare a 5-ng/µL working solution of psoralen–biotin in H_2O.

3. Add 40 µL of psoralen–biotin working solution to each 150 µL of DNA. Centrifuge the plate briefly and shake it on an orbital shaker at 800 rpm for approximately 5 minutes at room temperature.

4. Bind the psoralen–biotin to the DNA by UV cross-linking at 365 nm. The total dose should be 8800 mJ/cm^2 at 365 nm. (For example, with the Stratagene 1800 Crosslinker, this requires 45 minutes with the plate at a height of 7.9 cm from the benchtop.)

    It will be important to do some preliminary optimization experiments with the setup in your laboratory. Too little cross-linking will lead to an inadequate amount of captured DNA (this can be determined using PicoGreen binding; see Protocol 7, p. 172). Too much cross-linking will lead to poor protein expression even if there is enough DNA printed (this can be measured using an anti-GST signal).

5. Add 20 µL of 3 M sodium acetate (pH 5.5) to each well. Add 120 µL of isopropanol. Vortex the plate for 2 minutes at low speed.

6. Centrifuge the plate at 4000 rpm (maximum speed) for 20 minutes at room temperature.

7. Discard the supernatant.

8. Wash each pellet with 100 µL of room-temperature 80% ethanol.

    The wash is important. It helps to ensure removal of excess unincorporated linker.

9. Vortex the plate for 2 minutes at low speed.

10. Centrifuge the plate for 20 minutes at room temperature.

11. Discard the supernatant.

12. Invert the plate onto paper towels on the bench. Leave it there for 10 minutes.

    Do not dry the pellets for too long or they will be difficult to redissolve.

13. Prepare enough Master mix for two 96-well plates.

Streptavidin	10 mg
Deionized H$_2$O	2.9 mL
Polyclonal anti-GST antibody	30 µL
80 mM BS3 linker (in DMSO)	75 µL

> Add the linker just before using to avoid excess cross-linking. Final concentrations will be polyclonal GST antibody (1:100, 50 µg/mL), streptavidin (3 mg/mL), and BS3 linker (2 mM).

14. Add 15 µL of Master mix to each well, and shake the plate for 15 minutes.

15. Transfer 10 µL of each redissolved sample to the wells of a 384-well array plate.

> It is important that, prior to this step, the map of where the clones from the 96-well plate will reside in the 384-well plate is defined. Maximum attention is needed for this transfer to be accurate.

16. Prepare GST registration spots for the protein detection step (Protocol 6): Dilute 0.6 mg/mL GST at 1:20 in H$_2$O. Add BS3 linker to a final concentration of 2 mM.

> The importance of registration spots cannot be overstated. Particularly on high-density arrays, it is essential to have landmarks that indicate the geography of the slide. The registration samples should be added to the 384-well plate in positions that will yield a useful pattern on the final array. This will vary from experiment to experiment.

> We use GST for registration spots because the most commonly used NAPPA vectors add GST to the protein as the capture tag. The anti-GST antibody used to display captured protein will also light up the purified protein. If a well-quantified amount of pure protein is printed, it can also be used as a standard for estimating the protein yield of the expressed proteins. Other NAPPA vectors use the FLAG tag, the myc tag, and the GFP tag among others. Purified versions of these proteins or peptides should be substituted for GST at this step.

17. Prepare whole mouse IgG antibody registration spots for the protein interaction experiments (Experiment 11, Protocols 1 and 2): Dilute stock whole mouse IgG antibody at 1:20 in H$_2$O. Add BS3 linker to a final concentration of 2 mM.

> Protein interaction experiments are typically visualized with a monoclonal antibody directed at the query protein (or its tag). A horseradish peroxidase–coupled secondary antimouse IgG is used to detect the monoclonal antibody. Thus, the whole mouse IgG prepared here, once printed, will be detected by the secondary antibody and can act as registration spots for interaction experiments. If the primary antibody is derived from a species other than mouse, then use IgG from that species for the registration spots. For example, if the array is probed with human serum as the primary antibody (e.g., for a biomarker study), then human IgG would make good registration spots because the secondary will presumably be an anti-human IgG antibody.

18. Centrifuge the plate briefly to collect the samples at the bottom of the wells.

19. Array the samples at room temperature and with humidity at 45–60%.

> A detailed description of the printing programs is beyond the scope of this chapter, because each one depends on the arraying instrument used and the experimental design. With the chemistries described here, signal crossover will begin to become significant as the density exceeds 3000 features per standard 1" x 3" microscope slide, using 300-µm pins.

20. Place the printed slides in a humidified chamber. (We place the slides on an elevated platform in a covered tray with a layer of water on the bottom [Fig. 11].) Cover the chamber with foil to block the light, and place the chamber overnight at 4°C.

21. Remove the slides from the humidified chamber. Note: Open the lid carefully to avoid dripping condensate onto the printed slides.

22. Air-dry the slides at room temperature by placing them carefully (printed side up) onto paper towels. Store them at room temperature, protected from light and humidity, in the metal racks within the Lock & Lock storage boxes.

> Slides using the chemistries and surfaces described above can be stored under anhydrous conditions for several months without evident loss of signal.

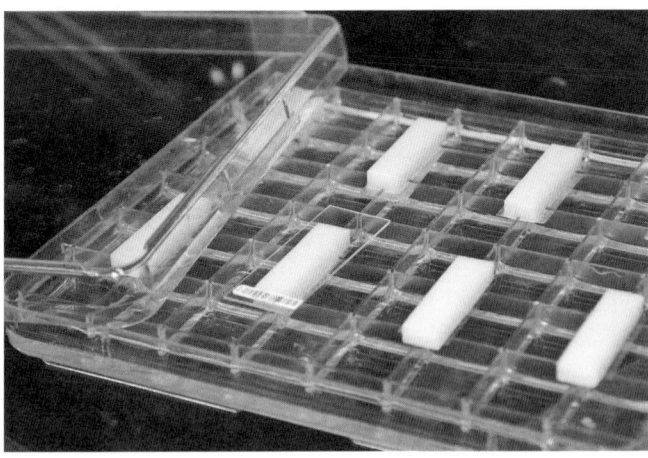

FIGURE 11. Humidification chamber. By adding water to the bioassay dishes with the 48 sector inserts and covering, a convenient humidification chamber can be assembled as shown here.

PROTOCOL 5

Expressing Proteins on NAPPA Slides

MATERIALS

Reagents

Blocking solution
   We use SuperBlock (Pierce, 37535) when the detection antibodies are monoclonals and 5% milk in phosphate-buffered saline (PBS) with 0.2% Tween 20 (i.e., "milk") when the antibodies are in serum.
Cell-free expression system, rabbit reticulocyte lysate (one tube per four slides; Promega, L4610)
H_2O, DEPC (diethylpyrocarbonate)-treated (Ambion, 9906)
PBS (pH 7.4)
RNaseOUT (Invitrogen, 10777-019)

Equipment

HybriWell sealing system (1 per slide; Grace Bio-Labs, HBW75)
Programmable chilling incubator, with leveled shelves
Rocking shaker
Slide incubators set at 15°C and 30°C
   See Step 7.

PROCEDURE

1. Block the slides with approximately 30 mL of SuperBlock or milk in the lid of a pipette tip box (or similar sized box; this works well for four slides). Make certain that the slides are submerged in the blocking solution. Incubate the slides for approximately 60 minutes on a rocking shaker at room temperature or overnight at 4°C in the cold room.

2. Rinse the slides with distilled deionized H_2O. Dry them gently with filtered compressed air.

3. Prepare the IVTT mix. Each slide requires 100 µL of the IVTT mix.

   a. Place the reticulocyte lysate on ice.

   b. Add the following components to the reticulocyte lysate. This will make enough for four slides.

TNT buffer (part of Promega kit)	4 μL
T7 polymerase (part of Promega kit)	2 μL
–Met (part of Promega kit)	1 μL
–Leu or –Cys (part of Promega kit)	1 μL
RNaseOUT (optional)	2 μL
DEPC-treated H$_2$O	40 μL

Do not vortex the tube! Instead, gently pipette the mixture up and down to mix.

4. Apply a HybriWell gasket to each slide. Use the wooden stick to rub along the adhesive to make sure it is well sealed everywhere. Apply the HybriWell to the same side of the slide as the printed features and align the gasket carefully so that the printed features are centered (Fig. 12).

5. Add IVTT mix to the HybriWell port that is opposite the bar code or specimen label. Both ports must be open to add the mix. Pipette the mix in slowly; it is okay if it beads up temporarily at the inlet end. As needed, gently massage the HybriWell along the inside edges of the adhesive to draw in the IVTT mix with capillary action. Ensure that the IVTT mix covers all of the printed area of the array and that there are no bubbles.

> This step is a tricky one to master and often requires practice. It may be useful to prepare some practice slides and a 10–20% glycerol/water mix with dye to practice this step before using a real sample. The practice slides are not a perfect representation of the real thing, which tends to be more difficult because of the surface chemistries of the arrays, but if the practice arrays are not mastered, it will be very difficult to get the real arrays to work well.

6. Seal both ports with the small round port seals (Fig. 12).

> It is essential to seal the ports after adding the mix to prevent drying and wicking of the signal.

7. Incubate the slides for 90 minutes at 30°C for protein expression, followed by 30 minutes at 15°C to allow the protein to be captured.

> Temperature precision during protein expression is critical; even a few degrees difference will reduce the yield.

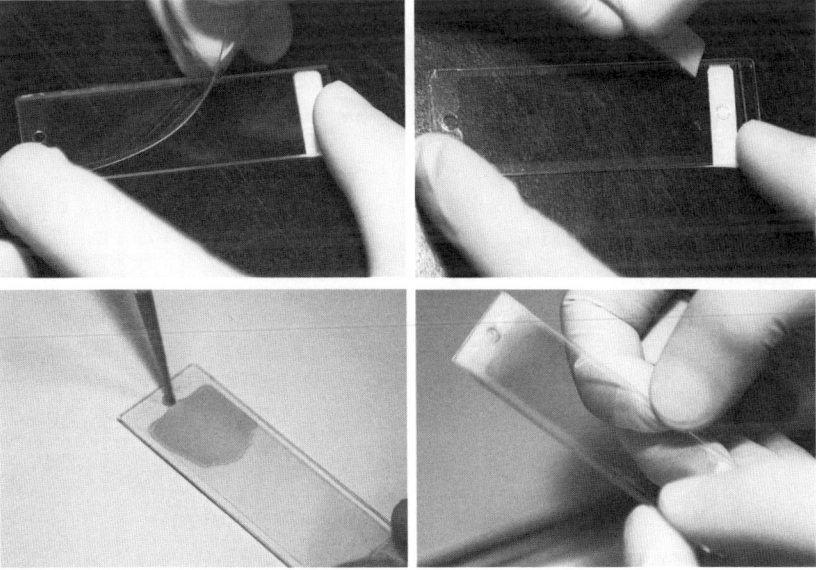

FIGURE 12. Protein expression cassette assembly for NAPPA. Apply HybriWell to slide carefully, ensuring that the printed part of the array is facing into the chamber and will fit entirely between the adhesive edges. Use the flat edge of the wooden stick to seal the adhesive along the slide, ensuring a good seal all around. After checking that both ports of the chamber are unobstructed, start adding the expression mix to one port at a consistent flow rate. Capillary action will draw the liquid into the chamber. More liquid can be drawn in by massaging the edges of the chamber (just inside the adhesive) unidirectionally in the direction of the flow.

8. Remove the HybriWell.
9. Wash the slides, in a pipette box, for 3 minutes with ~30 mL of milk.
10. Discard the wash solution and repeat Step 9 twice more.
11. Block the slides with approximately 30 mL of milk on a rocking shaker, either overnight at 4°C or at room temperature for 1 hour.

PROTOCOL 6

Detecting Proteins on NAPPA Slides

MATERIALS

Reagents

Blocking solution
   We use SuperBlock (Pierce, 37535) when the detection antibodies are monoclonals and 5% milk in PBS with 0.2% Tween 20 (i.e., "milk") when the antibodies are in serum.
PBS (pH 7.4)
Primary antibody (e.g., mouse anti-GST [Cell Signaling Technology, 2624] or mouse anti-HA [hemagglutinin antigen])
Secondary antibody, HRP(horseradish peroxidase)-conjugated anti-mouse (Amersham, NA931)
TSA cyanine-3 tyramide reagent pack (PerkinElmer, SAT704B001EA)

Equipment

Arrayed slides (from Protocol 5)
Coverslips, 24 × 50 (will need three per slide)
Pipette box
Scanner (e.g., PerkinElmer ScanArray 5000; Cy3 excitation 550 nm, emission 570 nm; Cy5 excitation 649 nm, emission 670 nm)

PROCEDURE

1. Prepare antibody solutions in SuperBlock. The dilutions will depend on the antibodies used and the specific experiment. Most conditions are close to those used for western blots. Store the diluted antibodies at 4°C.

   Typical experiments include a primary antibody (e.g., antiquery protein in interaction experiments or patient serum for immune response detections) and a marker-linked secondary antibody (e.g, HRP-conjugated anti-mouse IgG or Cy5-conjugated anti-human IgG). It is important to test all antibodies against the arrays ahead of time to ensure that they do not cross-react with any of the chemistries on the surface (i.e., causing all features on the array to respond). Where possible, test antibodies from several vendors to find one that has the best signal-to-noise ratio.

2. For HRP-conjugated antibodies, prepare the TSA stock solution.

   a. Add 150 µL of DMSO to dried TSA (from TSA kit).
   b. Vortex to dissolve.
   c. Keep this 100x stock solution at 4°C.

      The TSA reagent is a tyramide molecule linked to a label (e.g., Cy3, Cy5, or biotin), which is activated by HRP to form a free radical. The highly reactive, labeled free radical favors attachment to tyrosine moieties, where it forms a covalent bond. Thus, as the reaction continues, the label molecules continue to accumulate at any feature that has local HRP activity.

3. Add 150 µL of primary antibody (e.g., mouse anti-GST or mouse anti-HA) to the nonlabel or nonspecimen end of the slide. Apply a coverslip.

4. Incubate the slides for 60 minutes at room temperature.

5. Wash the slides with milk three times for approximately 5 minutes each. Drain the milk after each wash.

6. Add 150 µL of secondary antibody (e.g., HRP-conjugated anti-mouse) to the nonlabel or nonspecimen end of the slide. Apply a coverslip.

7. Incubate the slides for 1 hour at room temperature.

8. Wash the slides with PBS (pH 7.4) three times for approximately 5 minutes each.

9. Quickly rinse the slides with distilled, deionized H_2O.

    Before applying TSA solution in Step 10, ensure that the slides are neither too wet nor too dry. (If they are too wet, it will dilute the TSA. If they are too dry, the solution will not distribute evenly.)

10. Using the Amplification Diluent (from TSA kit), dilute the TSA stock solution 1:100 to a final volume that will allow 600 µL of TSA mix to be used for each slide. Although this volume is in excess, it makes processing easier when working with many slides. Add 600 µL to each slide and place a coverslip on the slide. Incubate the slide for 10 minutes.

11. Rinse the slide in distilled deionized H_2O. Dry it with filtered compressed air.

12. Scan the slide (Fig. 13).

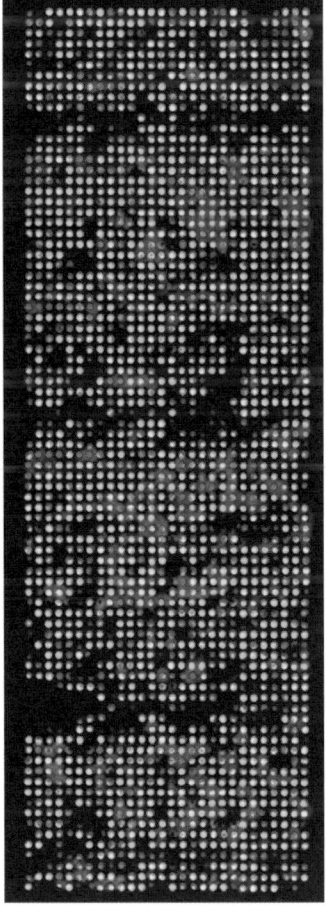

FIGURE 13. Protein levels on NAPPA array as indicated by anti-GST antibody. False color image indicating the levels of proteins on the array (*white* [saturated] > *red* > *yellow* > *green* > *blue*).

PROTOCOL 7

Detecting DNA on NAPPA Slides

MATERIALS

Reagents

PBS (pH 7.4)
PicoGreen stock solution (Invitrogen, P7581)
SuperBlock (Pierce, 37535)
TE (pH 8.0)

Equipment

Arrayed slides
Pipette box
Rocking shaker
Scanner (e.g., PerkinElmer ScanArray 5000)

PROCEDURE

1. Prepare the PicoGreen stock solution by adding 200 µL of TE to one vial (100 µL) of PicoGreen.

   The manufacturer recommends preparing the PicoGreen solution in a plastic container rather than glass, because the reagent may adsorb to glass surfaces. Protect the working solution from light by covering it with foil or placing it in the dark, because the reagent is susceptible to photodegradation. For best results, this solution should be used within a few hours of its preparation.

2. Prepare the PicoGreen working solution by making a 1:600 dilution of PicoGreen stock solution in SuperBlock.

3. Place one to four arrayed slides in the bottom of a box (an empty pipette tip box is commonly used). Block the slides with SuperBlock on a rocking shaker for 1 hour at room temperature.

   We prefer to submerge the slides in a small box with the reagent, but 24 x 50-mm coverslips offer a much more economical way to carry out these steps, if desired.

4. Add 20 mL of PicoGreen working solution to ensure that all of the slides are completely immersed (20 mL is sufficient for incubation in a pipette tip box). Shake on a rocking shaker for 5 minutes at room temperature.

5. Wash the slides with PBS (pH 7.4) three times for approximately 5 minutes each.

6. Quickly rinse the slides with distilled, deionized H_2O. Dry them with filtered compressed air.

7. Scan the slides (Fig. 14).

FIGURE 14. DNA levels on NAPPA as indicated by PicoGreen binding.

REFERENCES

MacBeath G. and Schreiber S.L. 2000. Printing proteins as microarrays for high-throughput function determination. *Science* **289:** 1760–1763.

Newman J.R.S. and Keating A.E. 2003. Comprehensive identification of human bZIP interactions with coiled-coil arrays. *Science* **300:** 2097–2101.

Ptacek J., Devgan G., Michaud G., Zhu H., Zhu X., Fasolo J., Guo H., Jona G., Breitkreutz A., Sopko R., et al. 2005. Global analysis of protein phosphorylation in yeast. *Nature* **438:** 679–684.

Zhu H., Bilgin M., Bangham R., Hall D., Casamayor A., Bertone P., Lan N., Jansen R., Bidlingmaier S., Houfek T., et al. 2001. Global analysis of protein activities using proteome chips. *Science* **293:** 2101–2105.

Zhu H., Hu S., Jona G., Zhu X., Kreiswirth N., Willey B.M., Mazzulli T., Liu G., Song Q., Chen P., et al. 2006. Severe acute respiratory syndrome diagnostics using a coronavirus protein microarray. *Proc. Natl. Acad. Sci.* **103:** 4011–4016.

EXPERIMENT 11

Using the Nucleic Acid Programmable Protein Array (NAPPA) for Identifying Protein–Protein Interactions

The identification of protein interactions using NAPPA (Nucleic Acid Programmable Protein Array) can be accomplished by either of two general schemas. The first probes an expressed NAPPA slide with a purified protein of interest (the query protein) and looks for interactors. Signals of these interactions can be detected either by directly labeling the query protein or by using a labeled antibody to the protein or to a tag on the protein. This approach works well when there is access to the purified protein, and it has the advantage that users can test query protein binding with and without posttranslational modifications or under a variety of binding conditions.

The second schema entails coexpressing the query protein on the NAPPA slide at the same time that all the target proteins are expressed. In contrast to the target proteins, which are printed on the array and have an epitope tag that facilitates protein capture directly to the surface, the gene for the query protein is added to the cell-free protein expression mix. The query protein either lacks a tag (if a good antibody to the query protein is available) or has a unique epitope tag intended for use during detection. Although fewer binding conditions can be tested in this schema compared with the one in which purified protein is used, this method has several advantages.

1. *Universality.* All that is needed is a copy DNA (cDNA) clone for the query gene, so this approach can be used on a wide variety of genes.

2. *Sensitivity.* A number of known protein interactions that require large amounts of purified protein for detection are more readily detected by coexpression. Although anecdotal, this might indicate that protein–protein binding is facilitated when proteins fold at the same time.

3. *Flexibility.* It is easy to test many different epitope tags for protein detection and to test many fragments of protein for mapping studies.

Although query protein binding can be detected with antibodies to the query protein, often these are unavailable. Instead, an epitope tag can be used that has no cross-reactivity with the capture tag and its affinity reagent. Several features of a potential epitope tag need to be considered.

Size. One of the most critical points when choosing a tag is its size. Large epitope tags, such as the GST (glutathione S-transferase) tag (26 kD), could, theoretically, create steric hindrance, particularly around the ends of the protein where they are attached. (There is some evidence that large tags position protein farther from the array surface, thereby improving access to the protein; although, for query proteins, this is not a major consideration.) Some large, well-characterized tags, like GST, rarely affect protein function. Small peptides, such as FLAG, myc, and HA (hemagglutinin antigen) (i.e., peptides of 7–12 amino acids), also tend not to interfere with the protein to which

TABLE 1. Selected epitope tag detection antibodies for NAPPA

Epitope	Antibody	Source
FLAG	M2	Sigma-Aldrich (F3165)
HA	12CA5	Sigma-Aldrich (H9658)
myc	9E10	Sigma-Aldrich (M4439)
GST	26H1	Cell Signaling Technology (2624)
GFP	anti-GFP	Invitrogen (a11120)

NAPPA, Nucleic Acid Programmable Protein Array; HA, hemagglutinin antigen; GST, glutathione S-transferase; GFP, green fluorescent protein.

they are fused and, thus, work well as tags for query proteins. The objective is to choose a tag that enables the query protein to maintain its native characteristics, such as conformation, in order to maximize opportunities for true binding interactions to occur.

Linker sequences. The inclusion of a linker sequence between the epitope tag and the query protein may increase accessibility of the epitope for the tag antibody, thereby facilitating detection of the query protein.

Protein folding. Some tags actually enhance protein folding. For example, thioredoxins sometimes display this characteristic when they are fused to the amino terminus of a query protein.

Detection reagents. The value of a tag in query detection relies on the availability of quality detection reagents. Usually, these are in the form of antibodies to the tag. Not all antibodies work well for detection on protein microarrays, particularly if the antibody suffers from cross-reactivity to other epitopes. Therefore, it is essential to screen several antibodies to assess the most suitable one. Table 1 (also see Protocol 2) provides a list of commercially available tag antibodies that perform well for the detection of the corresponding query proteins on NAPPA slides.

Tag location. The steric characteristics of an epitope might depend on which end of the protein harbors the tag. Tags at the carboxyl terminus allow confirmation that full-length protein translation of the protein has occurred. Efficiency of translation appears to depend on the amino acids proximal to the start of translation. The use of tags at the amino terminus ensures that all proteins begin with the same amino acids. In protein expression and purification from bacteria, the use of amino-terminal tags gives more reliable expression and better yields than when the native start codons are used (Braun et al. 2002; Dyson 2004). In the end, it may be useful to try both configurations or to consider constructs that have different tags at each end.

PROTOCOL 1

Expression of NAPPA Slides: Coexpression of Query Protein

The amount of query protein that is transcribed and translated from the corresponding plasmid DNA depends on the amount of plasmid DNA used and the size of the protein of interest, among other factors. If too little query protein is expressed, there may be no detectable binding signal. Excessive amounts of protein expression may generate nonspecific background signals. Because the optimum amount of query plasmid varies with each query protein, it is essential to assess empirically the optimal amount of query protein DNA to add to a coexpression experiment. We have found it works well to "bracket the exposure" by trying three different amounts, typically ranging from 50 to 150 ng of DNA per slide.

MATERIALS

Reagents

Blocking solution
> We use SuperBlock (Pierce, 37535), 30 mL per box (up to four slides) when the detection antibodies are monoclonals and 5% milk in phosphate-buffered saline (PBS) with 0.2% Tween 20 (i.e., "milk") when the antibodies are in serum.

Cell-free expression system, rabbit reticulocyte lysate (one tube per four slides; Promega, L4610)
H_2O, DEPC (diethylpyrocarbonate)-treated (Ambion, 9906)
PBS (pH 7.4)
Query plasmid (made from a miniprep; requires 50–150 ng/slide)
(Optional) RNaseOUT (Invitrogen, 10777-019)

Equipment

Arrayed slides (from Experiment 10, Protocol 4)
HybriWell sealing system (one per slide; Grace, HBW75)
Incubator, programmable chilling with leveled shelves
Pipette box lid (or similar sized box; see Step 1)
Shaker, rocking style
Slide incubators set at 15°C and 30°C
> See Step 7.

PROCEDURE

1. Block the slides with approximately 30 mL of SuperBlock or milk placed in the lid of a pipette tip box (or similarly sized box that will comfortably hold up to four slides). Rock the slides on a rocking shaker for approximately 60 minutes at room temperature or overnight at 4°C in the cold room. The slides should remain submerged in the blocking solution.

2. Rinse the slides with distilled deionized H_2O. Dry them gently with filtered compressed air.

3. Prepare the in vitro transcription/translation (IVTT) mix. Each slide requires 100 µL of the IVTT mix (50 µL of lysate and 50 µL of the remaining components). This reaction is sufficient to probe one slide.

    > As there are 200 µL of reticulocyte lysate in the manufacturer's tube, and the lysate should not be frozen again once it has been thawed, one lysate tube can be used for up to four slides at a time.

    a. Quickly thaw a tube of the reticulocyte lysate and immediately place it on ice.

    b. Add the following components to the lysate to create the IVTT mix (volumes listed are per single slide, i.e., per 50 µL of reticulocyte lysate).

TNT buffer (part of Promega kit)	4 µL
T7 polymerase (part of Promega kit)	2 µL
–Met (part of Promega kit)	1 µL
–Leu or –Cys (part of Promega kit)	1 µL
RNaseOUT(optional)	2 µL
DEPC water	40 µL

    > If several slides are processed at a time, a Master mix of the reticulocyte lysate plus components should be prepared. After that, the mix will be aliquoted according to the number of slides.

178 EXPERIMENT 11

    c. Aliquot the relevant miniprep DNA for the corresponding query protein(s) (50–150 ng per slide) to a set of wells or tubes.

    d. Add 100 µL of the IVTT mix to each well or tube containing query DNA.

        Do not vortex the tube! Instead, gently pipette the solution up and down to mix.

4. Apply a HybriWell gasket to each slide. Use the wooden stick to rub along the adhesive to make sure it is well sealed everywhere. Apply the HybriWell to the same side of the slide as the printed features and align the gasket carefully so that the printed features are centered (see Experiment 10, Fig. 12).

5. Add IVTT mix to the HybriWell port that is opposite the bar code or specimen label. Both ports must be open to add the mix. Pipette the mix in slowly; it is okay if it beads up temporarily at the inlet end. As needed, gently massage the HybriWell along the inside edges of the adhesive as needed to draw in the IVTT mix with capillary action. Ensure that the IVTT mix covers all of the printed area of the array and that there are no bubbles.

    This step is a tricky one to master and often requires practice. It may be useful to prepare some practice slides and a 10–20% glycerol/water mix with dye to practice this step before using real sample. The practice slides are not a perfect representation of the real thing, which tends to be more difficult because of the surface chemistries of the arrays, but if the practice arrays are not mastered, it will be very difficult to get the real arrays to work well.

6. Seal both ports with the small round port seals (see Experiment 10, Fig. 12).

    It is essential to seal the ports after adding the mix to prevent drying and wicking of the signal.

7. Incubate the slides for 90 minutes at 30°C for protein expression, followed by 2 hours at 15°C to allow the protein to be captured and the query to bind to its immobilized interacting targets.

    Temperature precision during protein expression is critical; even a few degrees difference will reduce the yield.

    If it is necessary to more evenly distribute the coexpressed query protein over the slide, the incubation time can be extended as follows. After the incubation at 15°C, transfer the slides to a humidified chamber and incubate at 4°C for about 12 hours (overnight is convenient; the lower temperature prevents protein degradation).

8. Remove the HybriWell.

9. Wash the slides in a pipette tip box for 3 minutes with 30 mL of milk.

10. Discard the wash solution and repeat Step 9 twice more.

11. Block the slides with approximately 30 mL of milk on a rocking shaker, either overnight at 4°C or at room temperature for 1 hour.

PROTOCOL 2

Detection of Query Proteins on NAPPA Slides

MATERIALS

CAUTION: See Appendix 11 for appropriate handling of materials marked with <!>.

Reagents

Blocking solution
> We use SuperBlock (Pierce, 37535) when the detection antibodies are monoclonals and 5% milk in phosphate-buffered saline (PBS) with 0.2% Tween 20 (i.e., "milk") when the antibodies are in serum.

Dimethylsulfoxide (DMSO) <!>
Primary antibody (e.g., mouse anti-GST [Cell Signal Technology, 2624] or mouse anti-HA)
Secondary antibody, HRP (horseradish peroxidase)-conjugated anti-mouse (Amersham, NA931)
TSA cyanine-3 tyramide reagent pack (PerkinElmer, SAT704B001EA)
PBS (pH 7.4)

Equipment

Arrayed slides (from Step 11 of Protocol 1)
Coverslips, 24 × 50 (will need three per slide)
Pipette box
Scanner (e.g., PerkinElmer ScanArray 5000; Cy3 excitation 550 nm, emission 570 nm; Cy5 excitation 649 nm, emission 670 nm)

PROCEDURE

1. Prepare antibody solutions in blocking solution. The dilutions will depend on the antibodies used and the specific experiment. Most conditions are close to those used for western blots. Store the diluted antibodies at 4°C.

   > Typical experiments include a primary antibody (e.g., anti-query protein in interaction experiments or patient serum for immune response detections) and a marker-linked secondary antibody (e.g., HRP-conjugated anti-mouse IgG or Cy5-conjugated anti-human IgG). It is important to test all antibodies against the arrays ahead of time to ensure that they do not cross-react with any of the chemistries on the surface (i.e., causing all features on the array to respond). Where possible, test antibodies from several vendors to find one that has the best signal-to-noise ratio.

2. For HRP-conjugated antibodies, prepare the TSA stock solution.

   a. Add 150 µL of DMSO to dried TSA (from TSA kit)

   b. Vortex the mixture.

   c. Keep this 100x stock solution at 4°C.

   > The TSA reagent is a tyramide molecule linked to a label (e.g., Cy3, Cy5, or biotin), which is activated by HRP to form a free radical. The highly reactive, labeled free radical favors attachment to tyrosine moieties, where it forms a covalent bond. Thus, as the reaction continues, the label molecules continue to accumulate at any feature that has local HRP activity.

3. Add 150 µL of primary antibody (e.g., mouse anti-GST or mouse anti-HA) to the nonlabel or nonspecimen end of the slide. Apply a coverslip.

4. Incubate the slides for 60 minutes at room temperature.

5. Wash the slides with milk three times for approximately 5 minutes each. Drain the milk after each wash.

6. Add 150 µL of secondary antibody (e.g., HRP-conjugated anti-mouse) to the nonlabel or nonspecimen end of the slide. Apply a coverslip.

7. Incubate the slides for 1 hour at room temperature.

8. Wash the slides with PBS (pH 7.4) three times for approximately 5 minutes each.

9. Quickly rinse the slides with distilled, deionized H₂O.

   Before applying TSA solution in Step 10, ensure that the slides are neither too wet nor too dry. (If they are too wet, it will dilute the TSA. If they are too dry, the solution will not distribute evenly.)

10. Using the Amplification Diluent (from the TSA kit), dilute the TSA stock solution 1:100 to a final volume that will allow 600 µL of TSA mix to be used for each slide. Although this volume is in excess, it makes processing easier when working with many slides.

11. Add 600 µL to each slide and place a coverslip on the slide. Incubate the slide for 10 minutes.

12. After the incubation, rinse the slides with water and air-dry them.

13. Scan the slides (Fig. 1).

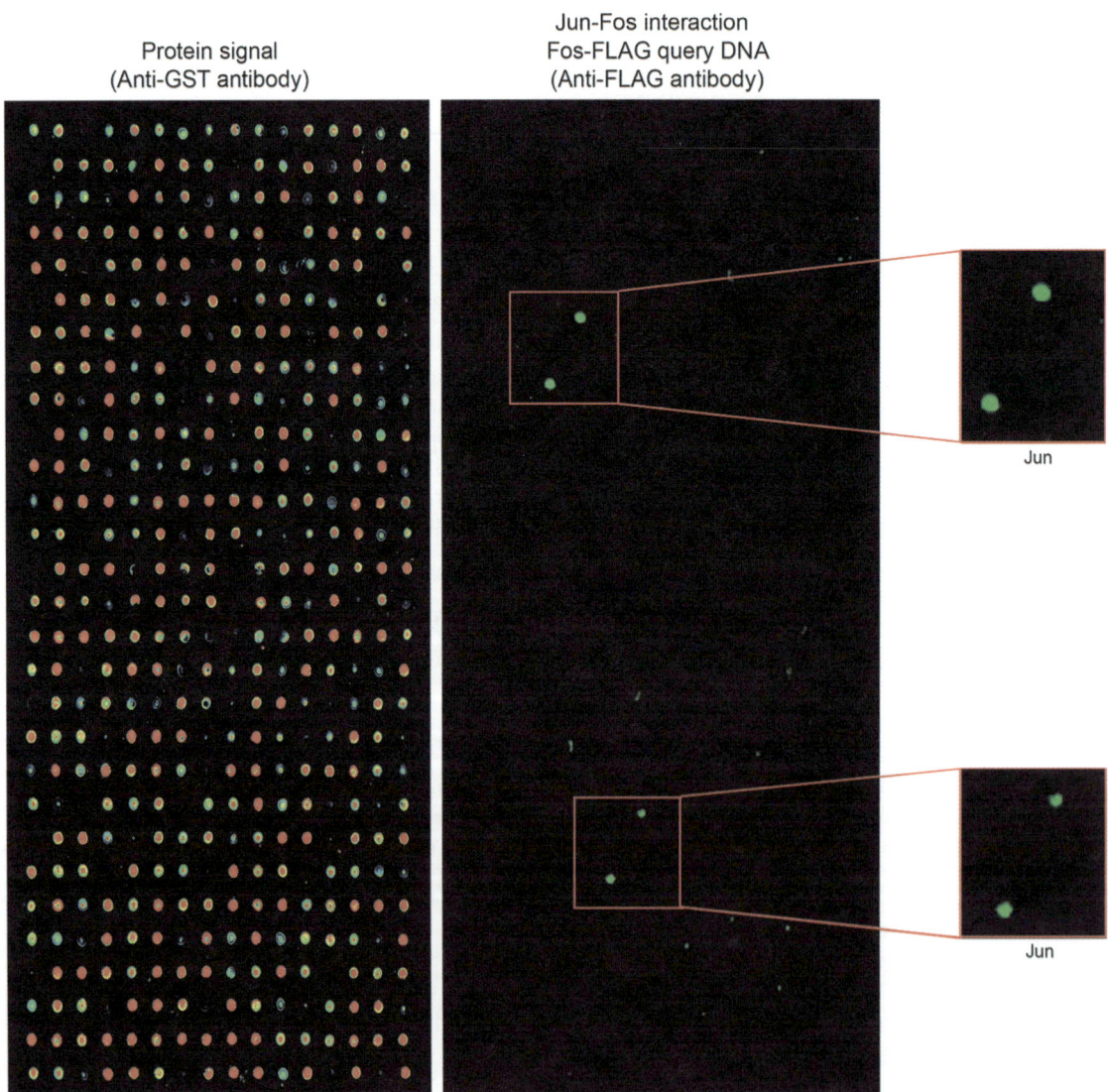

FIGURE 1. Identifying the interaction partner of Fos. Expression of target proteins on an array exhibiting more than 400 features was detected by anti-GST antibody (*left*). The Fos-FLAG query DNA was coexpressed on the same array and bound query protein detected by anti-FLAG antibody (*right*), identifying Jun as the interaction partner of Fos (*inset*). Images are actual data generated during CSHL Proteomics Course 2007 and are courtesy of Sanjeeva Srivastava, Harvard University.

REFERENCES

Braun P., Hu Y., Shen B., Halleck A., Koundinya M., Harlow E., and LaBaer J. 2002. Proteome-scale purification of human proteins from bacteria. *Proc. Natl. Acad. Sci.* **99:** 2654–2659.

Dyson M.R., Shadbolt S.P., Vincent K.J., Perera R.L., and McCafferty J. 2004. Production of soluble mammalian proteins in *Escherichia coli:* Identification of protein features that correlate with successful expression. *BMC Biotechnol.* **4:** 32.

APPENDIX 1

Setup and Demonstration of a Nanoelectrospray Ionization (nanoESI) Source and Tandem Mass Spectrometry (MS/MS)

An electrospray ionization source is assembled and an empty pulled microcapillary column is connected to the assembly for infusion of a peptide sample. The microcapillary column is positioned directly in front of the opening to the mass spectrometer. (See Experiment 4 for details on how to construct a microcapillary column.) For the CSHL Proteomics course, angiotensin I peptide is infused using a nanoESI source to demonstrate the fundamentals of electrospray ionization and MS/MS. The setup is also used to tune the mass spectrometer for the LC-MS/MS experiments (Experiments 4 and 6). After being tuned, the ion trap mass spectrometer is operated manually so students can perform precursor scans to detect and analyze precursor ions. Students then trap a selected ion and fragment the ion to demonstrate the principles of MS/MS.

MATERIALS

CAUTION: See Appendix 11 for appropriate handling of materials marked with <!>.

Reagents

Peptide or mixed peptide solution
Tune solution

Combine 500 µL methanol <!>, 494 µL Milli-Q H_2O, 1 µL formic acid <!>, 5 µL angiotensin (from a 20 pmol/µL stock; Sigma-Aldrich A9650)

Equipment

Camera
Fused silica capillary tubing, 100 µm ID x 365 µm OD (Polymicro Technologies)
Gold wire, 0.025 inch OD (Scientific Instrument Services W352)
Hamilton syringe, 250-µL (Hamilton 81175)
HPLC pump, quaternary (Agilent)
LTQ mass spectrometer (ThermoFisher)
Microcapillary column, unpacked, 100 µm ID x 365 µm OD (see Experiment 4)
MicroTight tubing sleeves, 380 µm, green (Upchurch Scientific F-185)

MicroTight ZDV adapter (Upchurch Scientific P-770)

Nanospray electrospray ionization source (James Hill Instruments)

    Each mass spectrometer vendor has its own conventional electrospray and nanospray ionization sources. For the CSHL courses, we use a custom nanospray ESI source because of its simplicity and easy access to the HPLC components for teaching the fundamentals of nanoLC-ESI-MS/MS.

PEEK double-winged nut with PEEK ferrule (Upchurch Scientific F-300, F-142x)

PEEK MicroTee (Upchurch Scientific P-775)

Teflon tubing, 1/6 inch OD (Upchurch Scientific 1513)

PROCEDURE

Assembling a NanoESI Source for Infusion of Angiotensin I

1. Construct a nanoESI source for infusion of a sample into the mass spectrometer (Fig. 1).

    a. Connect a 250-μL Hamilton syringe to a MicroTight adapter using a PEEK double-winged nut fitting, a PEEK ferrule, and a piece of Teflon tubing.

        The Teflon tubing acts as a sleeve to form a tight fit between the syringe and the MicroTight adapter.

    b. Insert a piece of 100 μm ID x 365 μm OD FSC (~20 inches long) into the MicroTight adapter with a MicroTight 380-μm green sleeve.

    c. Connect the infusion line to one arm of a PEEK MicroTee using a 380-μm green sleeve.

    d. Connect a piece of 0.025 inch OD gold wire through the center arm of the MicroTee and attach to the ESI voltage source.

    e. Connect a pulled, unpacked FSC column to the third arm of the MicroTee.

        See Experiment 4 for details on making a pulled FSC column.

    f. Disconnect the 250-μL syringe, fill with tune solution, and reattach the syringe to the infusion line. Install the syringe into the LTQ's syringe pump (Fig. 2).

2. Using the x-y-z manipulator, position the pulled-column needle tip in front of the mass spectrometer opening (Fig. 3). The needle tip should be between 1–5 mm from the opening of the mass spectrometer. Use a camera to assist in the positioning of the needle.

3. Open the "LTQ Tune" application.

    This allows the LTQ mass spectrometer to be operated manually.

4. Start the syringe pump flowing at 1 μL/min and check the flow from the needle tip.

    If no flow is detected, use a silica scribe to open the pulled FSC column tip as described in Experiment 4. The syringe pump allows you to infuse a sample solution into the ESI source for extended periods of time.

5. If available, load a previous tune file into LTQ Tune. If not, use any available tune file and manually change the capillary temperature to 150°C and spray voltage to 2.2 kV.

6. Turn on the source and start infusing the tune solution at 0.5 μL/min. Establish a stable 433 m/z ion by adjusting the position of the needle tip, the flow rate, or spray voltage.

    In the Tune Plus application, click on the "ON/Standby" button to turn on the MS detector. The MS detector begins scanning, the LTQ applies high voltage to the ESI source, and the LTQ detector shows a real time display in the Spectrum view. You should notice a strong, single peak at

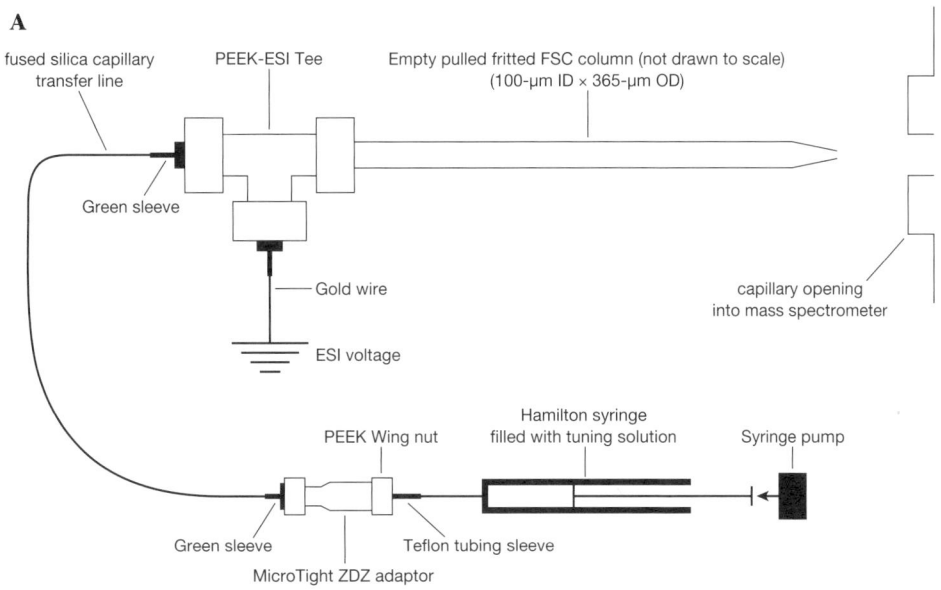
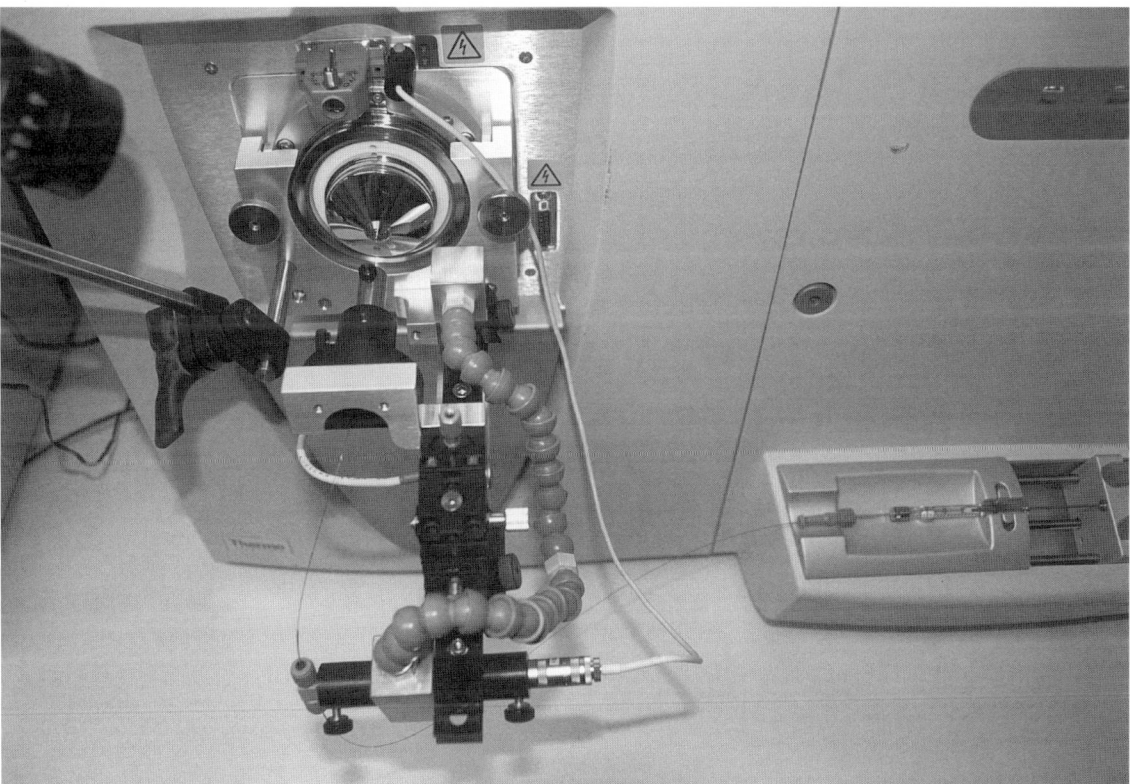

FIGURE 1. The nanoESI source for infusion. (A) Diagram of the components and FSC connections for infusing a sample to demonstrate the operation of the nanosprayESI source and ion trap mass spectrometer. (B) The LTQ linear ion trap mass spectrometer coupled with a nanosprayESI source for infusion of a sample.

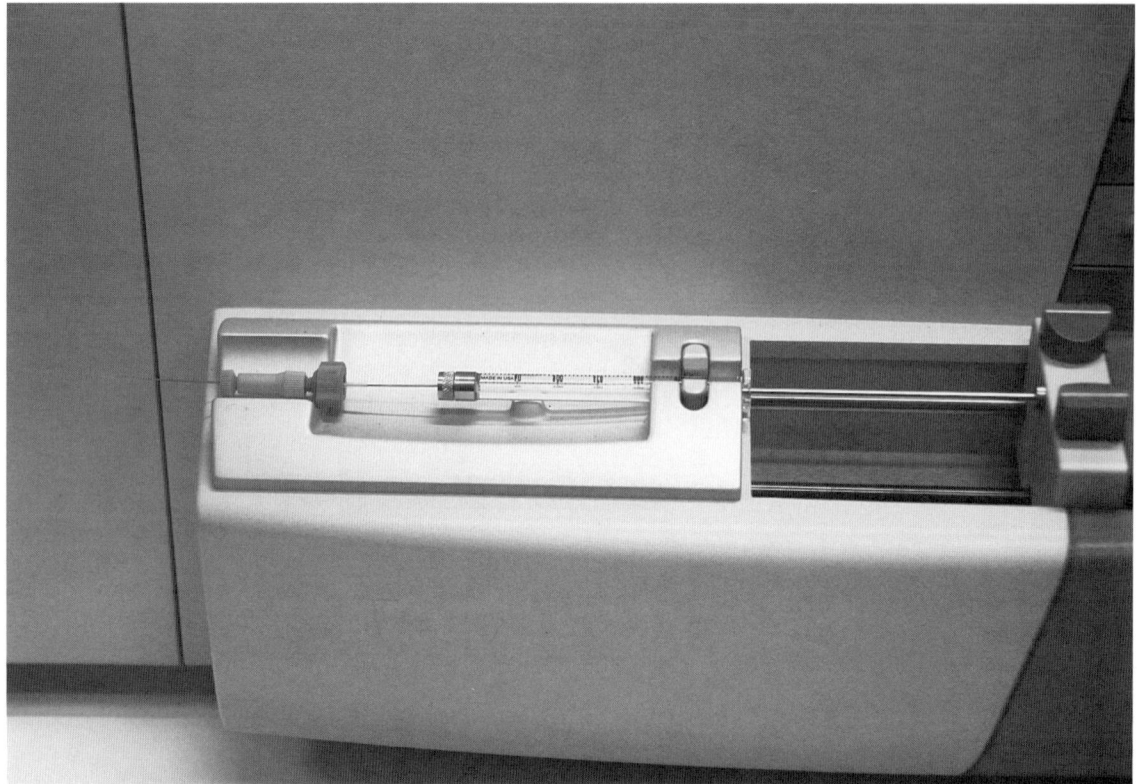

FIGURE 2. Setup of a Hamilton syringe and syringe pump for infusion of a sample into the mass spectrometer's ESI source.

433 m/z with little to no background ions. To establish a good signal, we initially infuse at 1 µL/min to establish the 433 signal and then adjust the flow rate to 0.5 µL/min. In the Scan Description Box, we select "Normal" to allow for a selection of mass ranges between m/z 300 to 2000. In the Scan Rate box, we select "Normal" scan rate. In the Scan Type box, we select "Full" for a full scan. In the Scan Time and Maximum Injection Time boxes, we enter "1 microscan" and "200ms," respectively.

7. Start the automated tune procedure using the 433 m/z ion.

   The LTQ is typically tuned using an automatic procedure. While automatic tuning is in progress, the LTQ MS detector displays various tests in the Spectrum and Graph views and various messages in the Status group box.

8. When an optimal file is created, save the tune file. This file will be used for subsequent LC-MS/MS experiments.

   Compare the new settings with the pretune settings to detect significant changes in the operating parameters. The process is repeated if the new tuning parameters decrease the signal quality.

Demonstrating the Fundamentals of Tandem Mass Spectrometry

1. Fill the 250-µL Hamilton syringe with a peptide or mixed peptide solution and connect the infusion line to the electrospray ionization source.

2. Start the LTQ syringe pump at a flow rate of 0.5 µL/min. Check to make sure that liquid is flowing from the column tip.

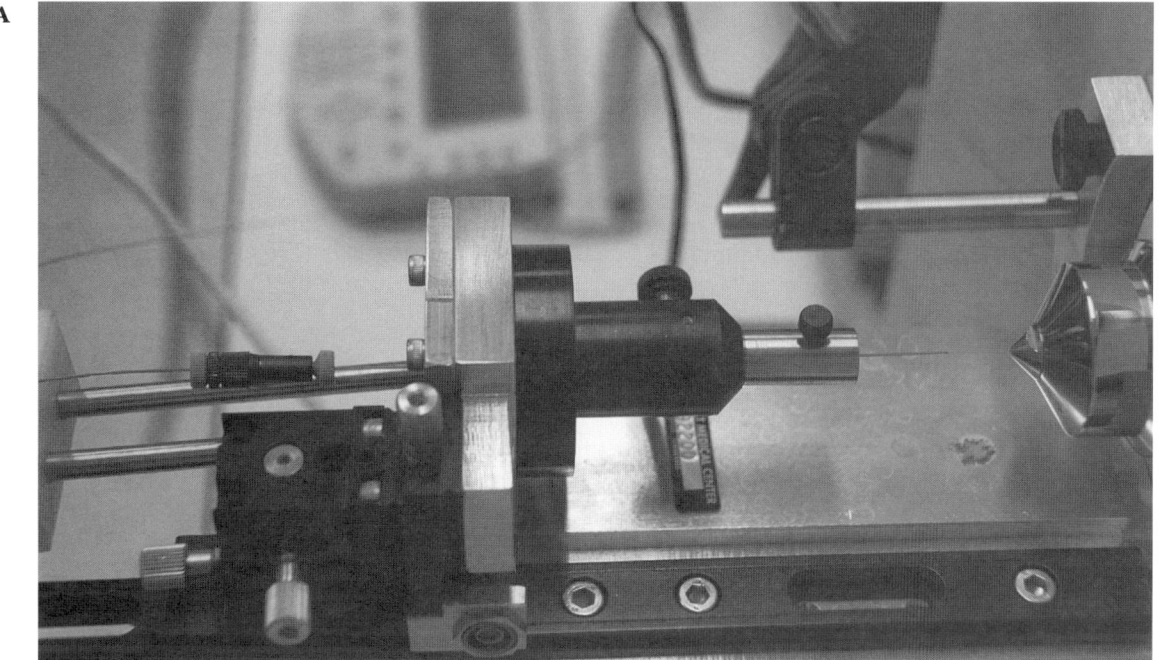

FIGURE 3. Positioning the ESI emitter relative to the capillary opening into the mass spectrometer. (A) nanoESI source and x-y-z micromanipulator used to hold and position the ESI emitter tip. (B) Position of the pulled fused silica capillary ESI emitter tip relative to the opening of the mass spectrometer's ion transfer tube for nanoESI.

3. Click on the "LTQ Tune" icon.

   This allows the mass spectrometer to be operated manually.

4. Select the appropriate tune file and then choose an *m/z* range of 300–2000 under "define scan."

5. Apply the voltage and check for peaks in the full MS scan.

   Once the voltage has been applied, a peak of interest should be evident (such as an angiotensin peak of 433 *m/z*) or if multiple peptides are being infused, several peaks should be detected. Toggle between centroid and profile mode to view the different data types. In the course, we demonstrate how to sum and average scans and how to save data to a file.

6. Once peaks are seen in the full MS scan, perform Zoom scans.

   When teaching, these scans are performed to train students how to determine the charge states of precursor ions.

7. After detecting precursor ions and determining the charge state in the full MS scan mode, try to fragment the ions of interest. In the course, the *m/z* 433 ion is trapped. Input the *m/z* value in the MS^n setting's Parent Mass (*m/z*) box, leave the normalized CID energy settings at "0," and click "apply."

   The selected ion (*m/z*) will now be trapped and detected as a single ion in the full scan. You should not see any fragmentation occurring at this point since collision induced dissociation (CID) energy is set at "0." This demonstration shows students how the ion trap is capable of trapping a single ion (*m/z*) while ejecting all other ions.

8. Increase the CID energy values until an optimal setting is found for fragmentation of your peptide.

   For demonstrating CID, CID energy (0, 10%, 15%, 25%, 35%) is added slowly to the trap ion to demonstrate the fragmentation of the ion and the process of MS/MS. Typically, 35% CID energy is optimal for our conditions in the course. This is also the default setting for the LTQ. In the course, we also demonstrate the ion trap's ability to do MS^n, by selecting one of the MS/MS ions and performing MS^3 on the fragment ion. The fragment ion is trapped and then the normalized collision energy is increased to 35% to fragment the ion.

9. Once your peptide has been successfully fragmented, return to the "define scan" box and clear the *m/z* value previously set for fragmentation.

   You should now see the full precursor MS scan.

After students have become familiar with the fundamentals of nanoESI and MS/MS, they are trained how to set up and program the mass spectrometer and microcapillary LC system to perform automated, data-dependent nanoLC-MS/MS or MudPIT experiments on unknown biological samples.

APPENDIX 2

Solution Protein Digest

Proteolytic digestion is critical for mass spectrometric sequencing because it generates peptides that have molecular weights within the mass range of the mass spectrometer. For tandem mass spectrometry, the digestion is typically performed using the protease trypsin, which cleaves at the carboxy-terminal side of lysine (K) and arginine (R) residues with the exception of K-P and R-P sites (Fig. 1).

There are numerous variations for "in solution" digestion of proteins. Because trypsin is a rather robust enzyme, tryptic digestions can be performed under various denaturing conditions (4 M urea, 2 M guanidine-HCl, 0.1% SDS, and >10% acetonitrile). Solubilization of proteins in 8 M urea and then diluting to 2–4 M urea before adding trypsin is commonly used to digest proteins that are difficult to solubilize. For most digestion protocols, cysteine residues are reduced and alkylated prior to digestion (Fig. 2). The reduction and alkylation of disulfide bonds helps to denature proteins, making their proteolytic sites more accessible to proteolysis.

Modified trypsin (TPCK-treated) is a serine endopeptidase prepared by treating trypsin with L-(tosylamido-2-phenyl) ethyl chloromethyl ketone (TPCK) to inactivate any remaining chymotryptic activity. TPCK acetylates the ε-amino groups of lysine residues to limit autolysis. Modified trypsin cleaves at K-P and R-P bonds at a much slower rate than other amino acid residues.

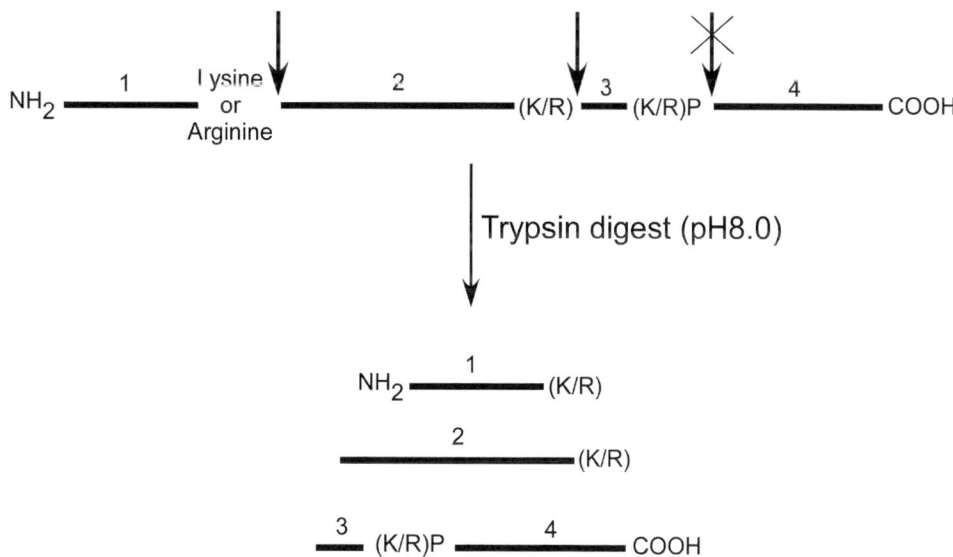

FIGURE 1. Cleavage sites for the site-specific endoprotease trypsin.

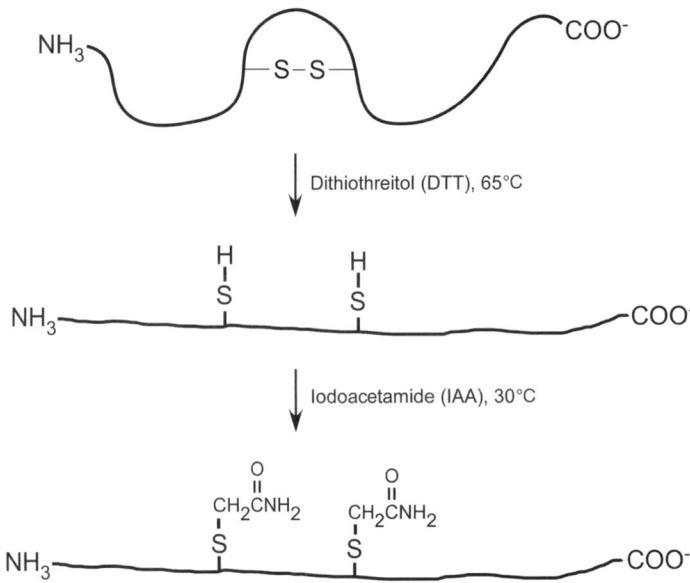

FIGURE 2. Reduction and alkylation of cysteine residues within proteins by DTT and IAA prior to proteolytic digestion.

MATERIALS

CAUTION: See Appendix 11 for appropriate handling of materials marked with <!>.

Reagents

Acetic acid <!> (0.5%)
10 mM Ammonium bicarbonate and 5% acetonitrile <!> solution
Dithiothreitol (DTT) <!> (50 mM) (Sigma-Aldrich, D-9163)
   Add 7.7 mg DTT (FW 154.2) to a 1.5-mL microcentrifuge tube. Add 1 mL of Milli-Q H_2O and mix. Prepare fresh each time before use.
Iodoacetamide (IAA) <!> (100 mM) (Sigma-Aldrich, I-1149)
   Add 18.5 mg IAA (FW 184) to a 1.5-mL microcentrifuge tube. Add 1 mL of Milli-Q H_2O. Store in the dark until ready to use. Prepare fresh.
Protein pellet
1 M Tris-Cl (pH 8.0) (Optional, see Steps 1 and 2)
Trypsin, modified sequencing-grade (Promega, V5111)

Equipment

Dry ice
Incubators set at 30°C, 37°C, and 65°C
0–14 pH paper

PROCEDURE

1. Dissolve the protein pellet in a solution of 100 mM ammonium bicarbonate and 5% acetonitrile.

   Trypsin activity is optimal at pH 8. Ammonium bicarbonate forms a mildly alkaline buffer when dissolved in water (pH 7.7) and is commonly used to buffer trypsin digestions. It is a simple volatile buffer that can be removed by lyophilization. Low concentrations of acetonitrile (<10%) faciliate trypsin proteolysis compared to ammonium bicarbonate buffer alone. For proteins that were detected by silver staining, resuspend the pellet in a volume of less than or equal to 20 µL. For Coomassie-detected proteins, resuspend in a volume from 20–100 µL. If the protein concentration is known, resuspend to a concentration of 10 µg/µL.

   The exact volume to resuspend a protein pellet should be considered arbitrary. Sometimes you will need to add more buffer to completely solubilize a pellet. For direct mass spectrometry analysis, avoid the use of detergents. For proteins already in solution, you typically only need to adjust to pH 8. We typically add 1/10 volume of 1 M Tris (pH 8.0) to adjust the pH. Trypsin is a rather robust enzyme.

2. Check the pH of the resuspended protein solution by spotting less than 1 µL onto the second square of a piece of 0–14 pH paper (Fig. 3A).

   The pH paper should turn blue indicating the pH is greater than 7.5. If the paper turns yellow, the sample is too acidic for trypsin to work. Continue adding 1/10 volume of 1 M Tris (pH 8.0) until the solution turns the pH paper blue (Fig. 3B).

3. Add 1/10 volume of 50 mM DTT. Incubate for 5 minutes at 65°C.

   The DTT reduces the disulfide bonds (see Fig. 2).

4. Add 1/10 volume of 100 mM iodoacetamide solution. Incubate for 30 minutes at 30°C in the dark.

   This step alkylates cysteine residues and prevents the formation of disulfide bonds. The alkylation step adds 57 Da to the mass of each cysteine residue. When searching the acquired spectra

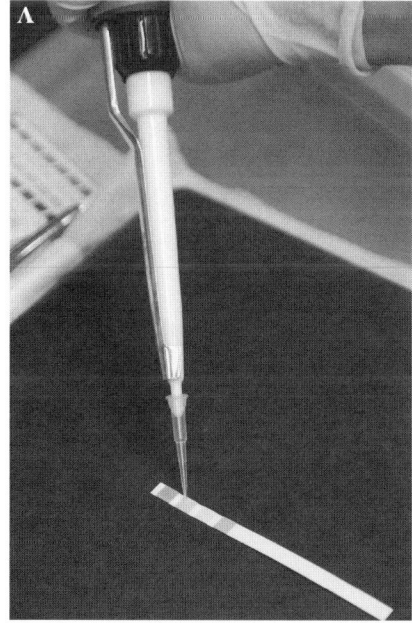
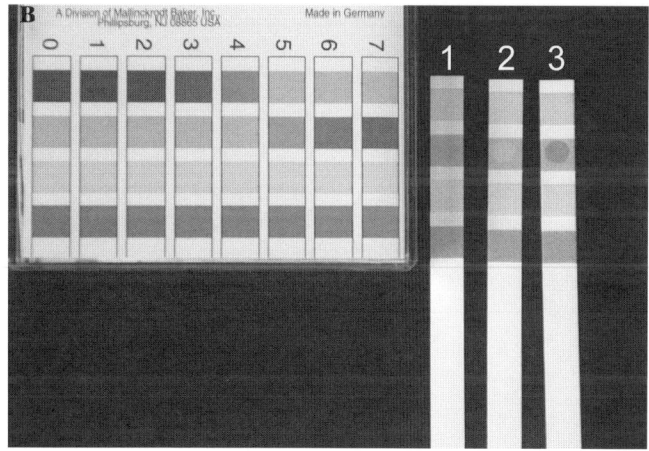

FIGURE 3. Checking the pH of a protein digest buffer prior to adding trypsin. (A) Spotting <1 µL of a protein digest solution onto the second square of 0–14 pH paper prior to adding trypsin. The square will turn yellow if the solution is acidic. When the pH is >7, the paper will turn blue. (B) Output from 0–14 pH paper showing pH of digestion buffer. The three pH paper strips to the left show (1) No solution spotted; (2) Acidic digestion solution; and (3) Alkaline digest solution.

against a protein database, all cysteine residues will have a fixed modification of 57 Da and the residue mass of cysteine will now be 160 Da (103 + 57 Da) (see Fig. 2).

5. Add modified sequencing-grade trypsin. Incubate overnight at 37°C.

   For the 20-μL digest of silver stain-detectable proteins, add 2 μL of **100 ng/μL** stock trypsin. For digesting Coomassie-detectable proteins, add 1–5 μL of **1 μg/μL** stock trypsin. If you know the total amount of protein in the sample, add trypsin to give a final substrate:trypsin ratio of 50:1.

6. Stop the reaction by adjusting the pH of the solution to less than pH 6 by acidifying the solution with 0.5% acetic acid (1–10 μL).

   Test the pH by spotting less than 1 μL aliquot of the sample onto the second square of pH 0–14 paper. The square should turn yellow indicating the digestion solution is acidic.

7. Snap-freeze the digest using dry ice and store the digested proteins at –20°C.

   Before mass spectrometry, centrifuge the digest at greater than 15,000g for 5 minutes. If a pellet is detected, transfer the supernatant to a fresh tube. Any solid particulates could clog the microcapillary HPLC column or nanospray ESI tip.

APPENDIX 3

In-gel Trypsin Digest of Gel-fractionated Proteins*

For a brief discussion on proteolytic digestion using trypsin, dithiothreitol, and iodoacetamide, see Appendix 2.

MATERIALS

CAUTION: See Appendix 11 for appropriate handling of materials marked with <!>.

Reagents

Use the purest chemicals available at all stages of sample preparation, including the gel casting. Gloves should be worn at all stages to minimize contamination by human epidermal proteins (keratins). The gloves should be powder-free to avoid contamination from talcum powder and dust.

Acetonitrile <!>

Ammonium bicarbonate (1 M)

Digestion solution

Prepare the digestion solution fresh and keep it on ice. Final concentrations are 100 mM ammonium bicarbonate, 0.5 mM $CaCl_2$ <!>, and 12.5 ng/µL modified sequencing-grade trypsin (Promega).

Dithiothreitol (DTT) <!> (50 mM) (Sigma-Aldrich, D-9163)

Add 7.7 mg DTT (FW 154.2) to a 1.5-mL microcentrifuge tube. Add 1 mL of Milli-Q H_2O and mix. Prepare fresh each time before use.

Formic acid <!>

Iodoacetamide (IAA) <!> (100 mM) (Sigma-Aldrich, I-1149)

Add 18.5 mg IAA (FW 184) to a 1.5-mL microcentrifuge tube. Add 1 mL of Milli-Q H_2O. Store in the dark until ready to use. Prepare fresh.

Protein sample(s)

Equipment

Dry ice
Glass plate
Incubators set at 30°C, 37°C, and 65°C
Mass spectrometer
Microcentrifuge tubes
Razor blade

---

*Adapted from Andrej Shevchenko and Matthias Mann's in-gel digestion protocol. An updated version of the protocol was released in 2006 with added information.

SDS-PAGE gel and electrophoresis apparatus
Spatula
Stains for gel-bound proteins (see note to Step 1)
Vacuum evaporator (e.g., SpeedVac)
Windex

> The Chait laboratory at The Rockefeller University routinely uses Windex to clean spatulas, razor blades, glass plates, and other items used for protein purification protocols and in-gel digests. They have found that Windex is highly efficient at removing contaminating proteins, such as keratins, that notoriously plague in-gel digests without leaving behind chemical contaminants.

PROCEDURE

1. Fractionate the protein sample(s) on an SDS-PAGE gel and stain.

   This protocol is applicable to both one- and two-dimensional polyacrylamide gels of different thickness, acrylamide concentration, and band (spot) size. The gels can be prepared by standard techniques using 0.1% SDS. Proteins are visualized by staining with Coomassie Brillant Blue R-250 or G-250, colloidal Coomassie, silver staining, SYPRO Ruby, or reverse staining (zinc-imidazole staining). If silver staining is used, select a protocol that is mass spectrometry-compatible (Shevchenko et al. 1996, 2006). Several mass spectrometry-compatible commercial stains are available, including SilverSNAP, GelCode Blue, and Imperial Protein Stain (all from Pierce). If silver, fluorescent, or reverse staining is used, then the extensive washing steps prior to reduction and alkylation can be omitted.

2. Photograph the gel and mark the bands to be identified.

3. Using a Windex-cleaned razor blade, spatula, and glass plate, excise the protein bands to be identified. Excise a gel piece of similar size from a nonprotein region of the gel and treat it identically to the protein band as a control throughout the in-gel digestion protocol and subsequent mass spectrometry analysis.

   Cut as close to the protein as possible to reduce the amount of background.

4. Cut the excised gel slice into 1-mm cubes and transfer the gel cubes to a 500-μL microcentrifuge tube.

5. Wash the gel pieces for 15 minutes with H_2O/acetonitrile mixture (1:1) for 15 minutes. All volumes should be at least twice the gel slice volume. Centrifuge quickly (1 sec) to transfer liquid from the sides of the tube to the bottom. Remove the liquid using a pipettor with a clean pipette tip.

6. Remove the liquid completely using a pipettor with a clean pipette tip.

   For Coomassie-stained bands, continue washing until the dye is extracted from the gel cubes. Most of the Coomassie should be removed.

7. Add acetonitrile to completely cover the gel pieces.

8. After the pieces have shrunk and turned sticky white, remove the acetonitrile.

9. Rehydrate in 100 mM ammonium bicarbonate.

10. After 5 minutes, add an equal volume of acetonitrile (to get a 1:1 ratio).

11. Incubate for 15 minutes and remove the liquid.

12. Dry the gel pieces completely in a vacuum evaporator.

    DO NOT apply heat to increase the drying rate.

13. Rehydrate the gel slices in a solution of 10 mM DTT and 100 mM ammonium bicarbonate. Add enough solution to completely cover the gel slices. Add more solution if it is absorbed by the gel pieces. Incubate for 45 minutes at 65°C.

    The 10 mM DTT, 100 mM ammonium bicarbonate solution reduces the disulfide bonds.

14. Remove the solution and quickly add an equal volume of 50 mM IAA, 100 mM ammonium bicarbonate buffer. Incubate for 30 minutes at 30°C in the dark.

    This step alkylates cysteine residues and prevents the formation of disulfide bonds. The alkylation step adds 57 Da to the mass of every cysteine residue. When searching the acquired spectra against a protein database, all cysteine residues will have a fixed modification of 57 Da and the residue mass of cysteine will now be 160 Da (103 + 57 Da).

15. Remove the IAA solution and wash the gel pieces as described in Steps 5–8.

16. Dry the gel pieces completely in a vacuum evaporator.

17. Add just enough digestion solution to cover the gel slices. Incubate the gel pieces on ice for 45 minutes. Add more digestion solution if all the initial solution is absorbed by the gel pieces.

18. Remove the excess digestion solution and add 5–20 µL of 100 mM ammonium bicarbonate to keep the gel pieces wet during enzymatic digestion.

19. Incubate overnight at 37°C.

20. Add a sufficient volume of 100 mM ammonium bicarbonate to cover the gel slices. Incubate for 15 minutes at room temperature.

21. Add an equal volume of acetonitrile and incubate for 15 minutes at room temperature.

22. Using a pipettor and a clean pipette tip, recover the supernatant and transfer it into a fresh 500-µL microcentrifuge tube.

23. Repeat the extraction two more times (Steps 20–22) but replace the 100 mM ammonium bicarbonate with 5% formic acid to stabilize the peptides. The recovered supernatants are pooled into the 500-µL microcentrifuge tube.

24. Freeze the pooled supernatant on dry ice.

25. Lyophilize the extracted peptides to near dryness. *DO NOT OVERDRY!* Ideally, stop when there is 1–2 µL of liquid left in the tube.

26. Resuspend the peptides in 2–20 µL of 5% formic acid.

27. Analyze using LC-MS/MS or MALDI TOF/TOF.

REFERENCE

Shevchenko A., Wilm M., Vorm O., and Mann M. 1996. Mass spectrometric sequencing of proteins silver-stained polyacrylamide gels. *Anal. Chem.* **68:** 850–858.

Shevchenko A., Tomas H., Havlis J., Olsen J.V., and Mann M. 2006. In-gel digestion for mass spectrometric characterization of proteins and proteomes. *Nat. Protoc.* **1:** 2856–2860.

APPENDIX 4

Trichloroacetic Acid (TCA) Precipitation of Proteins

MATERIALS

CAUTION: See Appendix 11 for appropriate handling of materials marked with <!>.

Reagents

Acetone <!>
Protein sample
100% Trichloroacetic acid (TCA) <!>, ice cold (Sigma-Aldrich, T-9159)
   Dissolve 2.2 g of TCA into 1 mL of dH_2O.

Equipment

Microcentrifuge
Vacuum evaporator (e.g., SpeedVac)

PROCEDURE

1. Add 0.11 volumes of ice-cold 100% TCA to the sample.

2. Place on ice for 10 minutes.

3. Add 500 µL of ice-cold 10% TCA to the sample.

4. Place on ice for 20 minutes.

5. Centrifuge at 20,000g for 30 minutes.

6. Carefully remove the supernatant.

   Aspiration of the supernatant with a pipette or a disposable transfer pipette works well. Avoid disturbing the protein pellet.

7. Add 500 µL of acetone. Very gently rock the tube once or twice to rinse the tube and pellet.

8. Centrifuge at 20,000g for 10 minutes. Very carefully remove the supernatant to avoid disturbing the protein pellet.

   Aspiration of the supernatant with a pipette or a disposable transfer pipette works well.

9. Dry the protein pellet in a vacuum evaporator (typically <5 min).

   The pellet is now ready for resuspension in a mass spectrometry compatible buffer.

APPENDIX 5

Monoisotopic and Immonium Ion Masses of Amino Acids*

Amino acid	3-letter code	1-letter code	Residue mass[a]	Immonium ion	Other low-mass ions
Glycine	Gly	G	57.021	30.034	
Alanine	Ala	A	71.037	44.050	
Serine	Ser	S	87.032	60.045	
Proline	Pro	P	97.053	70.066	
Valine	Val	V	99.068	72.081	41.0391, 55.0548, 69.0704
Threonine	Thr	T	101.048	74.061	
Cysteine	Cys	C	103.009	76.022	
Isoleucine	Ile	I	113.084	86.097	44.0500, 72.0449
Leucine	Leu	L	113.084	86.097	44.0500, 72.0449
Asparagine	Asn	N	114.043	87.056	70.0293
Aspartic acid	Asp	D	115.027	88.040	70.0293
Glutamine	Gln	Q	128.059	101.072	56.0500, 84.0449, 129.1028
Lysine	Lys	K	128.095	101.108	56.0500, 84.0813
Glutamic acid	Glu	E	129.043	102.056	84.0449
Methionine	Met	M	131.040	104.053	
Histidine	His	H	137.059	110.072	
Phenylalanine	Phe	F	147.068	120.081	91.0548
Arginine	Arg	R	156.101	129.114	70.0657, 100.0875, 112.0875
Tyrosine	Tyr	Y	163.063	136.076	91.0548, 107.0497
Tryptophan	Trp	W	186.079	159.092	77.039, 117.058, 130.066, 132.081
Carbamidomethyl Cys			160.031	133.044	
Oxidized Methionine			147.035	120.048	
Phosphoserine			166.998		
Phosphothreonine			181.014		
Phosphotyrosine			243.030		
Pyroglutamic acid			111.032		

[a] Amino acid – H_2O.

*Provided by Philip C. Andrews (*Department of Biological Chemistry, University of Michigan, Ann Arbor, Michigan 48109*)

APPENDIX 6

Dipeptide Masses of Amino Acids*

Dipeptide masses unprotonated—for use with gaps in de novo sequencing

		Gly G 57	Ala A 71	Ser S 87	Pro P 97	Val V 99	Thr T 101	Cys C 103	Leu/ Ile L/I 113	Asn N 114	Asp D 115	Gln/ Lys Q/K 128	Glu E 129	Met M 131	His H 137	Met- ox M-ox 147	Phe F 147	Arg R 156	Cys- cm C-cm 161	Tyr Y 163	Cys- am C-am 174	Trp W 186
Gly	57	114																				
Ala	71	128	142																			
Ser	87	144	158	174																		
Pro	97	154	168	184	194																	
Val	99	156	170	186	196	198																
Thr	101	158	172	188	198	200	202															
Cys	103	160	174	190	200	202	204	206														
Leu/Ile	113	170	184	200	210	212	214	216	226													
Asn	114	171	185	201	211	213	215	217	227	228												
Asp	115	172	186	202	212	214	216	218	228	229	230											
Gln/Lys	128	185	199	215	225	227	229	231	241	242	243	256										
Glu	129	186	200	216	226	228	230	232	242	243	244	257	258									
Met	131	188	202	218	228	230	232	234	244	245	246	259	260	262								
His	137	194	208	224	234	236	238	240	250	251	252	265	266	268	274							
Met-ox	147	204	218	234	244	246	248	250	260	261	262	275	276	278	284	294						
Phe	147	204	218	234	244	246	248	250	260	261	262	275	276	278	284	294	294					
Arg	156	213	227	243	253	255	257	259	269	270	271	284	285	287	293	303	303	312				
Cys-cm	161	218	232	248	258	260	262	264	274	275	276	289	290	292	298	308	308	317	322			
Tyr	163	220	234	250	260	262	264	266	276	277	278	291	292	294	300	310	310	319	324	326		
Cys-am	174	231	245	261	271	273	275	277	287	288	289	302	303	305	311	321	321	330	335	337	348	
Trp	186	243	257	273	283	285	287	289	299	300	301	314	315	317	323	333	333	342	347	349	360	372

*Provided by Philip C. Andrews (*Department of Biological Chemistry, University of Michigan, Ann Arbor, Michigan 48109*)

APPENDIX 7

LTQ Instrument Methods*

Instrument methods are provided for one-dimensional and multidimensional chromatography linked to peptide analysis by mass spectrometry.

One-dimensional Reversed-phase NanoLC-MS/MS Instrument Method

These instrument methods are used to set up and perform automated one-dimensional RP-LC-MS/MS (reversed-phase liquid chromatography/mass spectrometry/mass spectrometry) experiment runs. We typically set up the high-performance liquid chromatography (HPLC) and LTQ mass spectrometer to be controlled via the LTQ software, Xcalibur. However, you can choose to start and run these methods manually by programming the HPLC and MS individually and then simultaneously starting the HPLC pump and the LTQ mass spectrometer.

MATERIALS

CAUTION: See Appendix 11 for appropriate handling of materials marked with <!>.

Reagents

One-dimensional RP-LC buffers
    Buffer A: 5% acetonitrile <!>, 0.1% formic acid <!>
    Buffer B: 80% acetonitrile, 0.1% formic acid

Equipment

LTQ mass spectrometer (ThermoFisher)
Quaternary HPLC pump (Agilent, 1200)
    To perform in-line automated multidimensional protein identification technology (MudPIT) experiments, either a quaternary HPLC pump or two binary HPLC pumps are required. For the CSHL course, we use an Agilent 1200 quaternary pump along with an Agilent microplate autosampler interfaced to a ThermoFisher LTQ mass spectrometer for training students to set up and perform RP-nanoLC-MS/MS and MudPIT experiments. HPLC and mass spectrometers from other vendors can also be used for teaching.

---

*Jennifer L. Jennings (*Department of Microbiology and Immunology, Vanderbilt University School of Medicine, Nashville, Tennessee 37232*) contributed to the development of these methods.

PARAMETERS

HP1200 pump gradient method: 70min 1D LC-MS/MS gradient

0.00	0.200 mL/min
0.00	0% B
60.00	0.200 mL/min
60.00	40% B
70.00	0.200 mL/min
70.00	60% B
71.00	0% B
71.00	0.300 mL/min
90.00	0.300 mL/min
90.00	0% B

LTQ instrument method: Manual6_70min_meth.

Total time: 70 minutes
Number of segments: 1
Scan events: 6
Scan events
1. normal/ms/full/300–2000/positive/centroid
2. normal/msms/dependent scan/1st most intense ion/dynamic exclusion ON
3. normal/msms/dependent scan/2nd most intense ion/dynamic exclusion ON
4. normal/msms/dependent scan/3rd most intense ion/dynamic exclusion ON
5. normal/msms/dependent scan/4th most intense ion/dynamic exclusion ON
6. normal/msms/dependent scan/5th most intense ion/dynamic exclusion ON

MudPIT Instrument Methods*

These instrument methods are used for demonstrating how to set up and perform automated in-line two-dimensional SCX (strong cation exchange)-RP-LC-MS/MS experiment runs or MudPIT experiments. We set up the HPLC and LTQ mass spectrometer to be controlled via the LTQ software, Xcalibur. This requires an autosampler or other start device that is interfaced with the HPLC pumps and the mass spectrometer. For automated MudPIT experiments, the autosampler functions as the start instrument to trigger the HPLC gradients and LTQ instrument methods. The autosampler's Port 1 and Port 2 are connected to the LTQ mass spectrometer's peripheral control "start in" and the Agilent 1200 Quaternary pump's 9-pin remote port. For each of the 6 steps in the MudPIT run, the autosampler sends a start signal to both instruments to begin the instrument method and gradient.

MATERIALS

CAUTION: See Appendix 11 for appropriate handling of materials marked with <!>.

Reagents

MudPIT buffers
Buffer A: 5% acetonitrile <!>, 0.1% formic acid <!>
Buffer B: 80% acetonitrile, 0.1% formic acid
Buffer C: 500 mM ammonium acetate <!>, 5% acetonitrile, 0.1% formic acid

*Contributed by Michael P. Washburn (*Stowers Institute for Medical Research, Kansas City, Missouri 64110*).

Equipment

LTQ mass spectrometer (ThermoFisher)
Quaternary HPLC pump (Agilent, 1200)

To perform in-line automated MudPIT experiments, either a quaternary HPLC pump or two binary HPLC pumps are required. For the CSHL course, we use an Agilent 1200 quaternary pump along with an Agilent microplate autosampler interfaced to a ThermoFisher LTQ mass spectrometer for training students to set up and perform RP-nanoLC-MS/MS and MudPIT experiments. HPLC and mass spectrometers from other vendors can also be used for teaching.

PARAMETERS

VU6_MudPIT_000.meth

Autosampler Method

Time	Port	Status
0.00	1	Off
0.00	2	Off
0.01	1	On
0.01	2	On
0.05	1	Off
0.05	2	Off

HPLC Gradient

Time	Flow	Composition	
0.0	0.200	A = 100%	B = 0%
16.00	0.200	A = 60%	B = 40%
17.00	0.200	A = 0%	B = 100%
20.00	0.200	A = 0%	B = 100%

LTQ Instrument Method

20-minute method
6 scan events
current tune file

Scan events
1. normal/ms/full/300–2000/positive/centroid
2. normal/msms/dependent scan/1st most intense ion/dynamic exclusion ON
3. normal/msms/dependent scan/2nd most intense ion/dynamic exclusion ON
4. normal/msms/dependent scan/3rd most intense ion/dynamic exclusion ON
5. normal/msms/dependent scan/4th most intense ion/dynamic exclusion ON
6. normal/msms/dependent scan/5th most intense ion/dynamic exclusion ON

VU6_MudPIT_015.meth

Autosampler Method

Time	Port	Status
0.00	1	Off
0.00	2	Off
0.01	1	On
0.01	2	On
0.05	1	Off
0.05	2	Off

HPLC Gradient

Time	Flow	Composition		
0.00	0.200	A = 100%	B = 0%	C = 0%
3.00	0.200	A = 100%	B = 0%	C = 0%
3.10	0.200	A = 85%	B = 0%	C = 15%
5.00	0.200	A = 85%	B = 0%	C = 15%
5.10	0.200	A = 100%	B = 0%	C = 0%
10.00	0.200	A = 100%	B = 0%	C = 0%
10.10	0.200	A = 100%	B = 0%	C = 0%
25.00	0.200	A = 85%	B = 15%	C = 0%
117.00	0.200	A = 55%	B = 45%	C = 0%

LTQ Instrument Method

117-minute method
6 scan events
current tune file

Scan events
1. normal/ms/full/300–2000/positive/centroid
2. normal/msms/dependent scan/1st most intense ion/dynamic exclusion ON
3. normal/msms/dependent scan/2nd most intense ion/dynamic exclusion ON
4. normal/msms/dependent scan/3rd most intense ion/dynamic exclusion ON
5. normal/msms/dependent scan/4th most intense ion/dynamic exclusion ON

VU6_MudPIT_030.meth

Autosampler Method

Time	Port	Status
0.00	1	Off
0.00	2	Off
0.01	1	On
0.01	2	On
0.05	1	Off
0.05	2	Off

HPLC Gradient

Time	Flow	Composition		
0.00	0.200	A = 100%	B = 0%	C = 0%
3.00	0.200	A = 100%	B = 0%	C = 0%
3.10	0.200	A = 70%	B = 0%	C = 30%
5.00	0.200	A = 70%	B = 0%	C = 30%
5.10	0.200	A = 100%	B = 0%	C = 0%
10.00	0.200	A = 100%	B = 0%	C = 0%
10.10	0.200	A = 100%	B = 0%	C = 0%
25.00	0.200	A = 85%	B = 15%	C = 0%
117.00	0.200	A = 55%	B = 45%	C = 0%

LTQ Instrument Method

117-minute method
6 scan events
current tune file

Scan events
1. normal/ms/full/300–2000/positive/centroid
2. normal/msms/dependent scan/1st most intense ion/dynamic exclusion ON
3. normal/msms/dependent scan/2nd most intense ion/dynamic exclusion ON
4. normal/msms/dependent scan/3rd most intense ion/dynamic exclusion ON
5. normal/msms/dependent scan/4th most intense ion/dynamic exclusion ON
6. normal/msms/dependent scan/5th most intense ion/dynamic exclusion ON

VU6_MudPIT_050.meth

Autosampler Method

Time	Port	Status
0.00	1	Off
0.00	2	Off
0.01	1	On
0.01	2	On
0.05	1	Off
0.05	2	Off

HPLC Gradient

Time	Flow	Composition		
0.00	0.200	A = 100%	B = 0%	C = 0%
3.00	0.200	A = 100%	B = 0%	C = 0%
3.10	0.200	A = 50%	B = 0%	C = 50%
5.00	0.200	A = 50%	B = 0%	C = 50%
5.10	0.200	A = 100%	B = 0%	C = 0%
10.00	0.200	A = 100%	B = 0%	C = 0%
10.10	0.200	A = 100%	B = 0%	C = 0%
25.00	0.200	A = 85%	B = 15%	C = 0%
117.00	0.200	A = 55%	B = 45%	C = 0%

LTQ Instrument Method

117-minute method
6 scan events
current tune file

Scan events
1. normal/ms/full/300–2000/positive/centroid
2. normal/msms/dependent scan/1st most intense ion/dynamic exclusion ON
3. normal/msms/dependent scan/2nd most intense ion/dynamic exclusion ON
4. normal/msms/dependent scan/3rd most intense ion/dynamic exclusion ON
5. normal/msms/dependent scan/4th most intense ion/dynamic exclusion ON
6. normal/msms/dependent scan/5th most intense ion/dynamic exclusion ON

VU6_MudPIT_070.meth

Autosampler Method

Time	Port	Status
0.00	1	Off
0.00	2	Off
0.01	1	On
0.01	2	On
0.05	1	Off
0.05	2	Off

HPLC Gradient

Time	Flow	Composition		
0.00	0.200	A = 100%	B = 0%	C = 0%
3.00	0.200	A = 100%	B = 0%	C = 0%
3.10	0.200	A = 30%	B = 0%	C = 70%
5.00	0.200	A = 30%	B = 0%	C = 70%
5.10	0.200	A = 100%	B = 0%	C = 0%
10.00	0.200	A = 100%	B = 0%	C = 0%
10.10	0.200	A = 100%	B = 0%	C = 0%
25.00	0.200	A = 85%	B = 15%	C = 0%
117.00	0.200	A = 55%	B = 45%	C = 0%

LTQ Instrument Method

117-minute method
6 scan events
current tune file

Scan events
1. normal/ms/full/300–2000/positive/centroid
2. normal/msms/dependent scan/1st most intense ion/dynamic exclusion ON
3. normal/msms/dependent scan/2nd most intense ion/dynamic exclusion ON
4. normal/msms/dependent scan/3rd most intense ion/dynamic exclusion ON
5. normal/msms/dependent scan/4th most intense ion/dynamic exclusion ON
6. normal/msms/dependent scan/5th most intense ion/dynamic exclusion ON

VU6_MudPIT_100.meth

Autosampler Method

Time	Port	Status
0.00	1	Off
0.00	2	Off
0.01	1	On
0.01	2	On
0.05	1	Off
0.05	2	Off

HPLC Gradient

Time	Flow	Composition		
0.00	0.200	A = 100%	B = 0%	C = 0%
2.00	0.200	A = 100%	B = 0%	C = 0%
2.10	0.200	A = 0%	B = 0%	C = 100%
22.00	0.200	A = 0%	B = 0%	C = 100%
22.10	0.200	A = 100%	B = 0%	C = 0%
27.00	0.200	A = 100%	B = 0%	C = 0%
27.10	0.200	A = 100%	B = 0%	C = 0%
37.00	0.200	A = 80%	B = 20%	C = 0%
85.00	0.200	A = 30%	B = 70%	C = 0%
90.00	0.200	A = 0%	B = 100%	C = 0%
90.10	0.200	A = 0%	B = 100%	C = 0%
95.00	0.200	A = 0%	B = 100%	C = 0%
95.10	0.200	A = 100%	B = 0%	C = 0%
97.00	0.200	A = 100%	B = 0%	C = 0%

LTQ Instrument Method

97-minute method
6 scan events
current tune file

Scan events
1. normal/ms/full/300–2000/positive/centroid
2. normal/msms/dependent scan/1st most intense ion/dynamic exclusion ON
3. normal/msms/dependent scan/2nd most intense ion/dynamic exclusion ON
4. normal/msms/dependent scan/3rd most intense ion/dynamic exclusion ON
5. normal/msms/dependent scan/4th most intense ion/dynamic exclusion ON
6. normal/msms/dependent scan/5th most intense ion/dynamic exclusion ON

APPENDIX 8

Off-line Desalting of Peptide Mixtures

Reversed-phase-trapping (desalting) cartridges are used for concentrating and desalting peptide samples prior to analysis by LC-MS and MALDI-MS. They provide a robust and efficient method for concentrating very dilute peptide solutions. Because large volumes of wash buffer can be used to wash away unbound salts and other contaminants, the trapping cartridges are frequently used to clean up peptide samples prior to mass spectrometry analysis.

MATERIALS

CAUTION: See Appendix 11 for appropriate handling of materials marked with <!>.

Reagents

Buffer A: 2% acetonitrile <!> and 0.1% TFA <!>
Buffer B: 95% acetonitrile and 0.1% TFA
Buffer C: 70% formic acid <!> and 30% isopropanol <!>
Formic acid
Peptide sample

Equipment

Desalting cartridges (Michrom BioResources)
> The reversed-phase-desalting cartridges are available in 2 µg (CapTrap), 20 µg (MicroTrap), and 200 µg (MacroTrap) sample capacities.

Dry ice
Manual trap holder kit (Michrom BioResources)
> A number of other vendors offer off-line desalting systems. Most notably, ZipTip (Millipore) is a popular system for rapidly desalting small volumes and sample quantities.

Microcentrifuge
Syringe, 50-µL (Hamilton)
Vacuum evaporator (e.g., SpeedVac, Savant)

PROCEDURE

1. Place the reversed-phase (RP) trapping cartridge into the trap holder (Fig. 1B). Connect the syringe port and the capillary fitting to the trap holder (Fig. 1A,C). Tighten the components "finger tight."

   Use the appropriate size RP cartridge for the amount of sample to be desalted. The bed volumes for the different size cartridges are 0.5 μL (CapTrap), 5 μL (MicroTrap), and 50 μL (MacroTrap). The traps are bi-directional.

2. Using a 50-μL syringe, inject 10 column volumes of Buffer B through the trap cartridge (Fig. 1D).

   This highly organic solvent removes any nonpolar contaminants. We inject 5 μL, 50 μL, and 500 μL for the CapTrap, MicroTrap, and MacroTrap cartridges, respectively.

   Important: Look for leaks and tighten the components of the trap holder where necessary.

3. Inject 10 column volumes of Buffer C through the trap.

   This extremely stringent solvent washes away any contaminants and removes any previous sample peptides.

4. Inject 10 column volumes of Buffer A through the trap.

   This step equilibrates the trap to ensure that all of the organic solvent is removed to allow sample peptide trapping. Additional injections of 2% acetonitrile/0.1% TFA before injecting the sample can be performed to ensure that the trap is equilibrated.

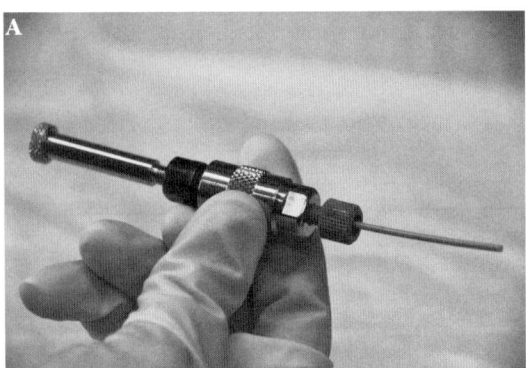
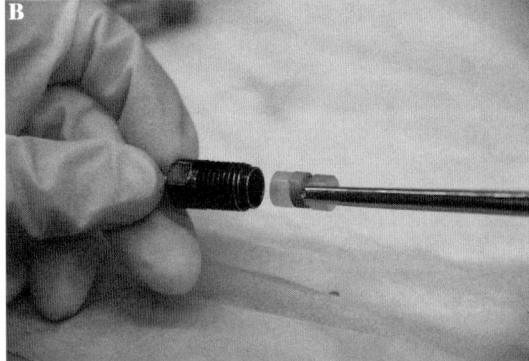
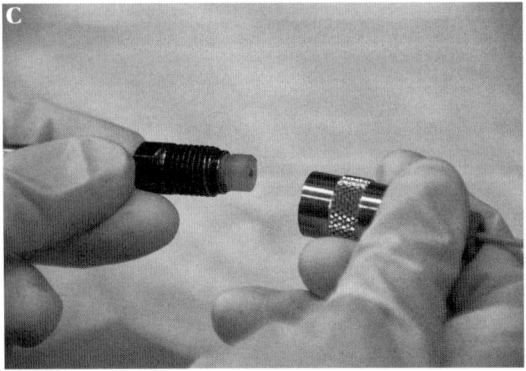
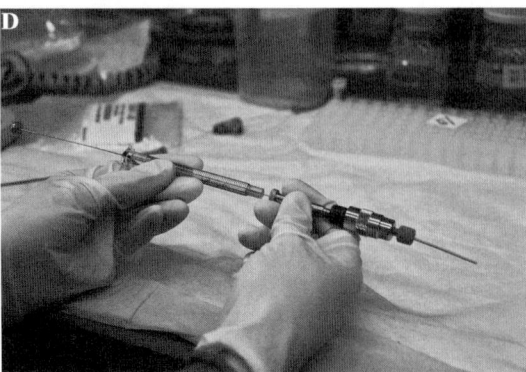

FIGURE 1. Reversed-phase (RP) trapping cartridge for both concentrating and desalting peptide samples prior to LC-MS and MALDI-MS analysis. (A) Desalting system for preparing peptide samples prior to LC-MS/MS analysis. The syringe port on the left allows mobile phase buffers and peptides solutions to be readily injected onto the trapping cartridge. (B) Using forceps to insert the RP-desalting cartridge into the holder. (C) Assembling the desalting holder with the RP cartridge installed. (D) Syringe used to inject mobile phases and peptide samples onto the desalting cartridge.

5. Pellet solid material that may plug the trapping cartridge by centrifuging the peptide sample for 1 minute at 14,000g.

6. Aspirate the sample into a clean 50-μL syringe.

7. Slowly inject the peptide sample through the equilibrated trap (Fig. 1D).

    During the injection of the sample, the flowthrough fraction can be collected in a microcentrifuge tube to recover peptides that fail to bind to the RP cartridge. The flow through can be reapplied to the cartridge to capture peptides that fail to bind initially.

8. Inject 10 column volumes of Buffer A through the trap.

    Additional washes with 2% acetonitrile/0.1% TFA may be necessary if the original peptide sample contains excessive amounts of salts, chaotropes, or detergents.

9. Inject 2 column volumes of Buffer B through the trap and collect the eluate into a labeled microcentrifuge tube.

    This solution is highly organic and elutes the peptides from the trap. This is the peptide fraction. To elute the peptides, we inject 1 μL, 10 μL, and 100 μL for the Captrap, Microtrap, and Macrotrap cartridges, respectively. Larger elution volumes can be used to assure maximum recovery of peptides from the trapping cartridges. For the MicroTrap, we typically elute the peptides with 50 μL of Buffer B.

10. Freeze the desalted peptide sample on dry ice and then lyophilize to a volume of 1–2 μL using a vacuum evaporator.

    Avoid excessive lyophilization, which can lead to loss of low abundance peptides.

11. Resuspend the peptides in 0.1% formic acid to a final volume of 1–10 μL.

12. Clean and equilibrate the microtrap with 20 column volumes washes of Buffer B, then 20 column volumes washes of Buffer C, and finally 20 column volumes washes of Buffer A.

13. Disassemble the trap holder and plug the ends of the trap to prevent the RP cartridge from drying out.

    The RP cartridge can be removed from the holder and stored indefinitely at room temperature in 0.1% formic acid.

APPENDIX 9

Preparing Competent Cells

It is important that the bacteria remain as viable as possible, so all resuspensions should be very gentle. It is useful to prechill pipettes, tips, dispensers, tubes, and plates that will be used to resuspend and aliquot bacteria.

MATERIALS

CAUTION: See Appendix 11 for appropriate handling of materials marked with <!>.

Reagents

DH5α cells
Dimethylsulfoxide (DMSO) <!>
LB (Luria broth) agar
LB medium
Transformation buffer, ice cold (final concentration in 1 liter)
  Combine

PIPES (piperazine-N, N'-bis[2-ethanesulfonic acid]) (10 mM)	3 g
$CaCl_2 \cdot 2H_2O$ <!> (15 mM)	2.2 g
KCl <!> (250 mM)	18.6 g

  Adjust the pH to 6.7–6.8 using KOH <!>.
  Add 10.9 g $MnCl_2 \cdot 4H_2O$ <!> (55 mM) and H_2O to bring the final volume to 1 liter.
  Filter the buffer using a 0.22-μm filter and store the buffer at 4°C.

Equipment

Liquid N_2 <!> or dry ice, crushed
  See Step 11.
Shaking incubator
  Because 20°C is sometimes lower than room temperature, results are more reliable if the incubator is kept in the cold room and heated to 20°C. This ensures a stable temperature.
Vials, prechilled
  See Step 10.

PROCEDURE

1. Streak DH5α cells on LB agar (no antibiotics) plates to isolate a colony (see Fig. 1).
2. Inoculate 5 mL of LB medium with a colony and grow overnight at 37°C with shaking.
3. Inoculate 500 mL of LB with the 5 mL preculture and grow for approximately 20 hours at 20°C (cold room incubator shaker set to 20°C at 300 rpm), until the OD_{600} = 0.5–0.7.
4. Centrifuge the culture at 3000 rpm for 10 minutes at 4°C.
5. Using a 25-mL pipette and working in the cold room, very gently resuspend the pellet in 80 mL of ice-cold Transformation buffer.
6. Incubate the tube on ice for 10 minutes.
7. Centrifuge the tube at 3000 rpm for 10 minutes at 4°C.
8. Using a 25-mL pipette and working in the cold room, very gently resuspend the pellet in 18.6 mL of ice-cold Transformation buffer.
9. Add 1.4 mL of DMSO (to a final concentration of 7%) and incubate the tube on ice for 10 minutes.
10. Gently distribute the competent cells into prechilled vials (microfuge tubes or PCR plates work well).

    For each 96-well plate to be used, you will need 2.5 mL in a 5-mL snap-top tube or 30 μL per well directly in the PCR plate.

11. Snap-freeze the cells on crushed dry ice or in liquid N_2.

    Avoid a dry ice/ethanol bath, because the ethanol may enter the tubes, which reduces the efficiency of transformation.

12. Store the competent cells at –80°C.

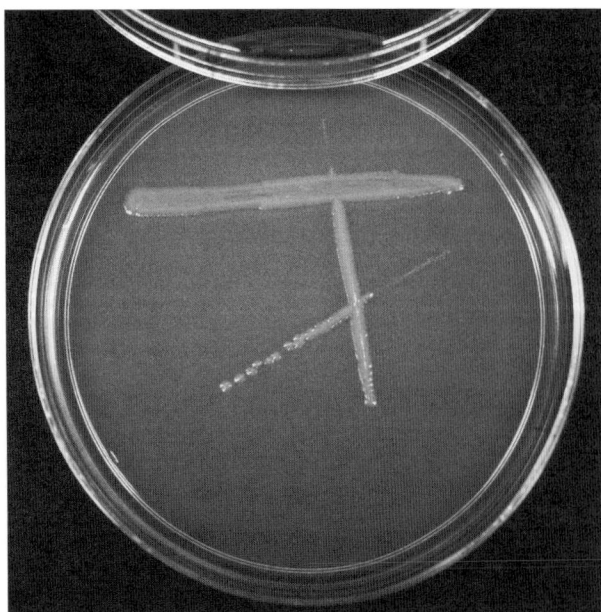

FIGURE 1. Method for streaking bacteria to get colonies. Use the pointed end of a square toothpick (not a round one) to make the first streak. Use one flat side of the other end of the toothpick to drag across the far end of the first streak creating the second streak. Turn the toothpick over and use the other flat side (the clean side) to drag across the second streak creating the third streak. Inevitably, there will be single colonies on either the second or third streak. (Thanks to M. Vidal.)

APPENDIX 10

DNA Quantification*

MATERIALS

CAUTION: See Appendix 11 for appropriate handling of materials marked with <!>.

Reagents

10x TNE buffer
    Tris (pH 7.5)   100 mM
    NaCl   1 M
    EDTA   10 mM
DNA samples to be quantified
DNA standards
Hoechst dye stock solution
    Dilute bisbenzimide <!> to 1 mg/mL in deionized H_2O. Store at –20°C.

Equipment

96-well flat-bottom black plate (Corning, 3915)
UV plate reader

PROCEDURE

1. Prepare a 1x TNE solution from the 10x stock.
2. Dilute the Hoechst dye reagent 1000-fold in 1x TNE.
3. Aliquot 95 µL of the diluted dye reagent into each well of a 96-well flat-bottom black plate.
4. Add 5 µL per well of DNA sample to be quantified.
5. Add a set of DNA standards of known concentrations (the concentration ranges should cover below and above the range of the expected concentration) to the appropriate wells.
6. Seal the plate and vortex it.
7. Centrifuge the plate briefly to bring everything down to the bottom of each well.
8. Read on a UV plate reader (absorbance 360 nm, emission 465 nm).
9. Draw a standard curve and calculate the sample concentration.

*Thanks to Frederick M. Boyce.

APPENDIX 11

Cautions

GENERAL CAUTIONS

Please note that the Cautions Appendix in this manual is not exhaustive. Readers should always consult individual manufacturers and other resources for current and specific product information. Chemicals and other materials discussed in text sections are not identified by the icon <!> used to indicate hazardous materials in the protocols. However, they may be hazardous to the user without special handling. Please consult your local safety office or the manufacturer's safety guidelines for further information.

The following general cautions should always be observed.

- **Become completely familiar with the properties of substances used before** beginning the procedure.
- **The absence of a warning** does not necessarily mean that the material is safe, because information may not always be complete or available.
- **If exposed** to toxic substances, contact your local safety office immediately for instructions.
- **Use proper disposal procedures** for all chemical, biological, and radioactive waste.
- **For specific guidelines on appropriate gloves**, consult your local safety office.
- **Handle concentrated acids and bases** with great care. Wear goggles and appropriate gloves. A face shield should be worn when handling large quantities.

  Do not mix strong acids with organic solvents because they may react. Sulfuric acid and nitric acid especially may react highly exothermically and cause fires and explosions.

  Do not mix strong bases with halogenated solvent as they may form reactive carbenes which can lead to explosions.

- **Handle and store pressurized gas containers** with caution because they may contain flammable, toxic, or corrosive gases; asphyxiants; or oxidizers. For proper procedures, consult the Material Safety Data Sheet that must be provided by your vendor.
- **Never pipette** solutions using mouth suction. This method is not sterile and can be dangerous. Always use a pipette aid or bulb.
- **Keep halogenated and nonhalogenated** solvents separately (e.g., mixing chloroform and acetone can cause unexpected reactions in the presence of bases). Halogenated solvents are organic solvents such as chloroform, dichloromethane, trichlorotrifluoroethane, and dichloroethane. Some nonhalogenated solvents are pentane, heptane, ethanol, methanol, benzene, toluene, N,N-dimethylformamide (DMF), dimethyl sulfoxide (DMSO), and acetonitrile.
- **Laser radiation**, visible or invisible, can cause severe damage to the eyes and skin. Take proper precautions to prevent exposure to direct and reflected beams. Always follow manufacturer's

safety guidelines and consult your local safety office. See caution below for more detailed information.

- **Flash lamps**, due to their light intensity, can be harmful to the eyes. They also may explode on occasion. Wear appropriate eye protection and follow the manufacturer's guidelines.
- **Photographic fixatives, developers, and photoresists** also contain chemicals that can be harmful. Handle them with care and follow manufacturer's directions.
- **Power supplies and electrophoresis equipment** pose serious fire hazard and electrical shock hazards if not used properly.
- **Microwave ovens and autoclaves** in the lab require certain precautions. Accidents have occurred involving their use (e.g., to melt agar or bacto-agar stored in bottles or to sterilize). If the screw top is not completely removed and there is not enough space for the steam to vent, the bottles can explode and cause severe injury when the containers are removed from the microwave or autoclave. Always completely remove bottle caps before microwaving or autoclaving. An alternative method for routine agarose gels that do not require sterile agar is to weigh out the agar and place the solution in a flask.
- **Ultrasonicators** use high frequency sound waves (16–100 kHz) for cell disruption and other purposes. This "ultrasound," conducted through air, does not pose a direct hazard to humans, but the associated high volumes of audible sound can cause a variety of effects, including headache, nausea, and tinnitus. Direct contact of the body with high-intensity ultrasound (not medical imaging equipment) should be avoided. Use appropriate ear protection and display signs on the door(s) of laboratories where the units are used.
- **Use extreme caution when handling cutting devices** such as microtome blades, scalpels, razor blades, or needles. Microtome blades are extremely sharp! Use care when sectioning. If unfamiliar with their use, have someone demonstrate proper procedures. For proper disposal, use the "sharps" disposal container in your lab. Discard used needles *unshielded*, with the syringe still attached. This prevents injuries (and possible infections; see Biological Safety) while manipulating used needles because many accidents occur while trying to replace the needle shield. Injuries may also be caused by broken Pasteur pipettes, coverslips, or slides.
- **Animal treatment:** Procedures for the humane treatment of animals must be observed at all times. Consult your local animal facility for guidelines. Animals, such as rats, are known to induce allergies and these allergies can increase in intensity with repeated exposure. Always wear a lab coat and gloves when handling the animal. If allergies to dander or saliva are known, wear a mask.

GENERAL PROPERTIES OF COMMON CHEMICALS

The hazardous materials list can be summarized in the following categories.
- Inorganic acids, such as hydrochloric, sulfuric, nitric, or phosphoric, are colorless liquids with stinging vapors. Avoid spills on skin or clothing. Spills should be diluted with large amounts of water. The concentrated forms of these acids can destroy paper, textiles, and skin as well as cause serious injury to the eyes.
- Inorganic bases such as sodium hydroxide are white solids that dissolve in water and under heat development. Concentrated solutions will slowly dissolve skin and even fingernails.
- Salts of heavy metals are usually colored, powdered solids which dissolve in water. Many of them are potent enzyme inhibitors and therefore toxic to humans and to the environment (e.g., fish and algae).

- Most organic solvents are flammable volatile liquids. Avoid breathing the vapors, which can cause nausea or dizziness. Also avoid skin contact.
- Other organic compounds, including organosulphur compounds such as mercaptoethanol or organic amines, can have very unpleasant odors. Others are highly reactive and should be handled with appropriate care.
- If improperly handled, dyes and their solutions can stain not only your sample, but also your skin and clothing. Some of them are also mutagenic (e.g., ethidium bromide), carcinogenic, and toxic.
- Nearly all names ending with "ase" (e.g., catalase, β-glucuronidase, or zymolyase) refer to enzymes. There are also other enzymes with nonsystematic names like pepsin. Many of them are provided by manufacturers in preparations containing buffering substances, etc. Be aware of the individual properties of materials contained in these substances.
- Toxic compounds are often used to manipulate cells. They can be dangerous and should be handled appropriately.
- Be aware that several of the compounds listed have not been thoroughly studied with respect to their toxicological properties. Handle each chemical with the appropriate respect. Although the toxic effects of a compound can be quantified (e.g., LD_{50} values), this is not possible for carcinogens or mutagens where one single exposure can have an effect. Also realize that dangers related to a given compound may also depend on its physical state (fine powder vs. large crystals/diethylether vs. glycerol/dry ice vs. carbon dioxide under pressure in a gas bomb). Anticipate under which circumstances during an experiment exposure is most likely to occur and how best to protect yourself and your environment.

HAZARDOUS MATERIALS

Acetic acid (concentrated) must be handled with great care. It may be harmful by inhalation, ingestion, or skin absorption. Wear appropriate gloves and goggles. Use in a chemical fume hood.

Acetic acid (glacial) is highly corrosive and must be handled with great care. It may be a carcinogen. Liquid and mist cause severe burns to all body tissues. It may be harmful by inhalation, ingestion, or skin absorption. Wear appropriate gloves and goggles and use in a chemical fume hood. Keep away from heat, sparks, and open flame.

Acetone causes eye and skin irritation and is irritating to mucous membranes and upper respiratory tract. Do not breathe the vapors. It is also extremely flammable. Wear appropriate gloves and safety glasses. Keep away from heat, sparks, and open flame.

Acetonitrile (Methyl cyanide) is very volatile and extremely flammable. It is an irritant and a chemical asphyxiant that can exert its effects by inhalation, ingestion, or skin absorption. Treat cases of severe exposure as cyanide poisoning. Wear appropriate gloves and safety glasses and use only in a chemical fume hood. Keep away from heat, sparks, and open flame.

AgNO₃, *see* **Silver nitrate**

Alanine is irritating to the eyes, skin, and respiratory system. It may be harmful by inhalation, ingestion, or skin absorption. Wear appropriate gloves and safety glasses.

Amino silane is an irritant and may cause severe corneal injury. It may also be harmful by inhalation, ingestion, and skin absorption. Wear appropriate gloves and safety goggles. Keep away from contact with water.

3-Aminopropyltriethoxysilane (TESPA), *see* **Silane**

Ammonium hydroxide, NH_4OH, is a solution of ammonia in water. It is caustic and should be handled with great care. As ammonia vapors escape from the solution, they are corrosive, toxic, and can be explosive. Use only with mechanical exhaust. Wear appropriate gloves and use only in a chemical fume hood.

Aspartic acid is a possible mutagen and poses a risk of irreversible effects. It may be harmful by inhala-

Note: In general, proprietary materials are not listed here. Kits and other commercial items as well as most anesthetics, dyes, fixatives, and stains are also not included. Anesthetics also require special care. Follow the manufacturer's safety guidelines that accompany these products.

tion, ingestion, or skin absorption. Wear appropriate gloves and safety glasses. Do not breathe the dust.

Bisbenzimide may be harmful by inhalation, ingestion, or skin absorption. Wear appropriate gloves and safety glasses and use in a chemical fume hood. Do not breathe the dust.

C_2H_4INO, *see* **Iodoacetamide**

$C_7H_7FO_2S$, *see* **Phenylmethylsulfonyl fluoride**

$CaCl_2$, *see* **Calcium chloride**

Calcium chloride, $CaCl_2$, is hygroscopic and may cause cardiac disturbances. It may be harmful by inhalation, ingestion, or skin absorption. Do not breathe the dust. Wear appropriate gloves and safety goggles.

CHAPS, *see* **3-[(3-Cholamidopropyl)dimethyl-ammonio]-1-propanesulfonate**

$CHCl_3$, *see* **Chloroform**

CH_3CH_2OH, *see* **Ethanol**

Chloroform, $CHCl_3$, is irritating to the skin, eyes, mucous membranes, and respiratory tract. It is a carcinogen and may damage the liver and kidneys. It is also volatile. Avoid breathing the vapors. Wear appropriate gloves and safety glasses and always use in a chemical fume hood.

3-[(3-Cholamidopropyl)dimethyl-ammonio]-1-propanesulfonate (CHAPS) is an irritant and may be harmful by inhalation, ingestion, or skin absorption. Wear appropriate gloves and safety glasses.

α-Cyano-4-hydroxycinnamic acid (HCCA) may cause cardiac disturbances. Chronic effects may be delayed. It may be harmful by inhalation, ingestion, or skin absorption. Wear appropriate gloves and safety glasses.

Digitonin may be fatal if inhaled, ingested, or absorbed through the skin. Wear appropriate gloves and safety glasses and use in a chemical fume hood.

Dimethyl sulfoxide (DMSO) may be harmful by inhalation or skin absorption. Wear appropriate gloves and safety glasses and use in a chemical fume hood. DMSO is also combustible. Store in a tightly closed container. Keep away from heat, sparks, and open flame.

Dithiothreitol (DTT) is a strong reducing agent that emits a foul odor. It may be harmful by inhalation, ingestion, or skin absorption. When working with the solid form or highly concentrated stocks, wear appropriate gloves and safety glasses and use in a chemical fume hood.

DMSO, *see* **Dimethyl sulfoxide**

DTT, *see* **Dithiothreitol**

Ethanol (EtOH), CH_3CH_2OH, is highly flammable and may be harmful by inhalation, ingestion, or skin absorption. Wear appropriate gloves and safety glasses. Keep away from heat, sparks, and open flame.

$FeCl_3$, *see* **Ferric chloride**

Ferric chloride, $FeCl_3$, may be harmful by inhalation, ingestion, or skin absorption. Wear appropriate gloves and safety glasses and use only in a chemical fume hood.

Formaldehyde, HCHO, is highly toxic and volatile. It is also a possible carcinogen. It is readily absorbed through the skin and is irritating or destructive to the skin, eyes, mucous membranes, and upper respiratory tract. Avoid breathing the vapors. Wear appropriate gloves and safety glasses and always use in a chemical fume hood. Keep away from heat, sparks, and open flame.

Formamide is teratogenic. The vapor is irritating to the eyes, skin, mucous membranes, and upper respiratory tract. It may be harmful by inhalation, ingestion, or skin absorption. Wear appropriate gloves and safety glasses and always use a chemical fume hood when working with concentrated solutions of formamide. Keep working solutions covered as much as possible.

Formic acid, HCOOH, is highly toxic and extremely destructive to tissue of the mucous membranes, upper respiratory tract, eyes, and skin. It may be harmful by inhalation, ingestion, or skin absorption. Wear appropriate gloves and safety glasses (or face shield) and use in a chemical fume hood.

Glacial acetic acid, *see* **Acetic acid (glacial)**

Guanidine hydrochloride is irritating to the mucous membranes, upper respiratory tract, skin, and eyes. It may be harmful by inhalation, ingestion, or skin absorption. Wear appropriate gloves and safety glasses. Avoid breathing the dust.

HCCA, *see* **α-Cyano-4-hydroxycinnamic acid**

HCHO, *see* **Formaldehyde**

HCl, *see* **Hydrochloric acid**

H_3COH, *see* **Methanol**

HCOOH, *see* **Formic acid**

$HOCH_2CH_2SH$, *see* **β-Mercaptoethanol (2-Mercaptoethanol)**

Hydrochloric acid, HCl, is volatile and may be fatal if inhaled, ingested, or absorbed through the skin. It is extremely destructive to mucous membranes, upper respiratory tract, eyes, and skin. Wear appropriate gloves and safety glasses and use with great care in a chemical fume hood. Wear goggles when handling large quantities.

Imidazole is corrosive and may be harmful by inhalation, ingestion, or skin absorption. Wear appropriate gloves and safety glasses and use in a chemical fume hood.

Iodoacetamide, C_2H_4INO, can alkylate amino groups in proteins and can therefore cause problems if the antigen is being purified for amino acid sequencing. It is toxic and harmful by inhalation, ingestion, or skin absorption. Wear appropriate gloves and safety glasses and use only in a chemical fume hood. Do not breathe the dust.

Isopropanol is flammable and irritating. It may be harmful by inhalation, ingestion, or skin absorption. Wear appropriate gloves and safety glasses. Do not breathe the vapor. Keep away from heat, sparks, and open flame.

KCl, see **Potassium chloride**

KOH, see **Potassium hydroxide**

Leupeptin (or its **hemisulfate**) may be harmful by inhalation, ingestion, or skin absorption. Wear appropriate gloves and safety glasses and use in a chemical fume hood.

Liquid nitrogen (LN_2) can cause severe damage due to extreme temperature. Handle frozen samples with extreme caution. Do not breathe the vapors. Seepage of liquid nitrogen into frozen vials can result in an exploding tube upon removal from liquid nitrogen. Use vials with O-rings when possible. Wear cryomitts and a face mask. Do not allow the liquid nitrogen to spill onto your clothes. Do not breathe the vapors.

LN_2, see **Liquid nitrogen**

Magnesium chloride, $MgCl_2$, may be harmful by inhalation, ingestion, or skin absorption. Wear appropriate gloves and safety glasses and use in a chemical fume hood.

Manganese chloride, $MnCl_2$, may be harmful by inhalation, ingestion, or skin absorption. Wear appropriate gloves and safety glasses and use in a chemical fume hood.

McOH, see **Methanol**

β-Mercaptoethanol (2-Mercaptoethanol), $HOCH_2CH_2SH$, may be fatal if inhaled or absorbed through the skin and is harmful if ingested. High concentrations are extremely destructive to the mucous membranes, upper respiratory tract, skin, and eyes. β-Mercaptoethanol has a very foul odor. Wear appropriate gloves and safety glasses and always use in a chemical fume hood.

MES, see **2-(N-Morpholino)ethanesulfonic acid**

Methanol, MeOH or H_3COH, is toxic and can cause blindness. It may be harmful by inhalation, ingestion, or skin absorption. Adequate ventilation is necessary to limit exposure to vapors. Avoid inhaling these vapors. Wear appropriate gloves and safety goggles and use only in a chemical fume hood.

$MgCl_2$, see **Magnesium chloride**

$MnCl_2$, see **Manganese chloride**

MOPS, see **3-(N-Morpholino)-propanesulfonic acid**

3-(N-Morpholino)-propanesulfonic acid (MOPS) may be harmful by inhalation, ingestion, or skin absorption. It is irritating to mucous membranes and upper respiratory tract. Wear appropriate gloves and safety glasses and use in a chemical fume hood.

2-[N-Morpholino]ethanesulfonic acid (MES) may be harmful by inhalation, ingestion, or skin absorption. Wear appropriate gloves and safety glasses.

Na_2CO_3, see **Sodium carbonate**

NaF, see **Sodium fluoride**

NaN_3, see **Sodium azide**

NaOH, see **Sodium hydroxide**

Na_3VO_4, see **Sodium orthovanadate**

NH_4OH, see **Ammonium hydroxide**

Periodic acid is a strong oxidizer. Contact with other material may cause fire. It is also corrosive and may be harmful by inhalation, ingestion, or skin absorption. Wear appropriate gloves and safety goggles.

Phenylmethylsulfonyl fluoride (PMSF), $C_7H_7FO_2S$, is a highly toxic cholinesterase inhibitor. It is extremely destructive to the mucous membranes of the respiratory tract, eyes, and skin. It may be fatal by inhalation, ingestion, or skin absorption. Wear appropriate gloves and safety glasses and always use in a chemical fume hood. In case of contact, immediately flush eyes or skin with copious amounts of water and discard contaminated clothing.

PMSF, see **Phenylmethylsulfonyl fluoride** ($C_7H_7FO_2S$)

Potassium chloride, KCl, may be harmful by inhalation, ingestion, or skin absorption. Wear appropriate gloves and safety glasses.

Potassium hydroxide, KOH, and **KOH/methanol**, are highly toxic and may be fatal if swallowed. It may be harmful by inhalation, ingestion, or skin absorption. Solutions are corrosive and can cause severe burns. It should be handled with great care. Wear appropriate gloves and safety goggles.

Psoralen is highly corrosive, especially to the eyes and skin and may be carcinogenic. It may be harmful by inhalation, ingestion, or skin absorption. Wear appropriate gloves and safety goggles. Do not breathe the dust.

SDS, see **Sodium dodecyl sulfate**

Silane is extremely flammable and corrosive. It may be harmful by inhalation, ingestion, or skin absorption. Keep away from heat, sparks, and open flame. The vapor is irritating to the eyes, skin, mucous membranes, and upper respiratory tract. Wear appropriate gloves and safety goggles and always use in a chemical fume hood.

Silver nitrate, $AgNO_3$, is a strong oxidizing agent and should be handled with care. It may be harmful by inhalation, ingestion, or skin absorption. Avoid contact with skin. Wear appropriate gloves and safety glasses. It can cause explosions upon contact with other materials.

Sodium azide, NaN$_3$, is highly poisonous. It blocks the cytochrome electron transport system. Solutions containing sodium azide should be clearly marked. It may be harmful by inhalation, ingestion, or skin absorption. Wear appropriate gloves and safety goggles and handle it with great care. Sodium azide is an oxidizing agent and should not be stored near flammable chemicals.

Sodium carbonate, Na$_2$CO$_3$, may be harmful by inhalation, ingestion, or skin absorption. Wear appropriate gloves and safety glasses and use in a chemical fume hood.

Sodium deoxycholate is irritating to mucous membranes and the respiratory tract and may be harmful by inhalation, ingestion, or skin absorption. Wear appropriate gloves and safety glasses when handling the powder. Do not breathe the dust.

Sodium dodecyl sulfate (SDS) is toxic, an irritant, and poses a risk of severe damage to the eyes. It may be harmful by inhalation, ingestion, or skin absorption. Wear appropriate gloves and safety goggles. Do not breathe the dust.

Sodium fluoride, NaF, is highly toxic and causes severe irritation. It may be fatal by inhalation, ingestion, or skin absorption. Wear appropriate gloves and safety glasses and use only in a chemical fume hood.

Sodium hydroxide, NaOH, and **solutions containing NaOH,** are highly toxic and caustic and should be handled with great care. Wear appropriate gloves and a face mask. All other concentrated bases should be handled in a similar manner.

Sodium orthovanadate, Na$_3$VO$_4$, may be harmful by inhalation, ingestion, or skin absorption. Wear appropriate gloves and safety glasses and use in a chemical fume hood.

SYPRO Orange/Red/Ruby contains **DMSO.** See **DMSO.**

TCA, *see* **Trichloroacetic acid**

TCEP, *see* **Tris-(carboxyethyl)phosphine hydrochloride**

TESPA (3-aminopropyltriethoxysilane), *see* **Silane**

TFA, *see* **Trifluoroacetic acid**

Thionyl chloride reacts violently with water (liberates toxic gas) and causes severe burns. It is highly toxic and harmful by inhalation, ingestion, or skin absorption. Wear appropriate gloves and safety glasses and use in a chemical fume hood. Do not breathe the vapor.

Thiourea may be carcinogenic and may be harmful by inhalation, ingestion, or skin absorption. Wear appropriate gloves and safety glasses and use in a chemical fume hood.

Trichloroacetic acid (TCA) is highly caustic. Wear appropriate gloves and safety goggles.

Triethylamine is highly toxic and flammable. It is extremely corrosive to the mucous membranes, upper respiratory tract, eyes, and skin. It may be harmful by inhalation, ingestion, or skin absorption. Wear appropriate gloves and safety glasses and use in a chemical fume hood. Keep away from heat, sparks, and open flame.

Trifluoroacetic acid (TFA) (concentrated) may be harmful by inhalation, ingestion, or skin absorption. Concentrated acids must be handled with great care. Decomposition causes toxic fumes. Wear appropriate gloves and a face mask and use in a chemical fume hood.

Tris (carboxyethyl) phosphine hydrochloride (TCEP) is corrosive to the mucous membranes, upper respiratory tract, eyes, and skin, and can cause burns. It may be harmful by inhalation, ingestion, or skin absorption. Wear appropriate gloves and safety glasses and use in a chemical fume hood. Do not breathe the vapor or mist.

Triton X-100 causes severe eye irritation and burns. It may be harmful by inhalation, ingestion, or skin absorption. Wear appropriate gloves and safety goggles. Do not breathe the vapor.

Trizol may be fatal if absorbed through the skin, inhaled, or swallowed. It can also cause severe burns. Wear appropriate gloves, safety goggles, protective clothing, and always use in a chemical fume hood. Rinse any areas of skin that come in contact with trizol with a large volume of water and wash with soap and water; do not use ethanol!

Trypsin may cause an allergic respiratory reaction. It may be harmful by inhalation, ingestion, or skin absorption. Do not breathe the dust. Wear appropriate gloves and safety goggles. Use with adequate ventilation.

Urea may be harmful by inhalation, ingestion, or skin absorption. Wear appropriate gloves and safety glasses.

Index

Amino acids
dipeptide masses, 201
monoisotopic and immonium ion masses, 199

Biotinylation, DNA in nucleic acid programmable protein array, 165–167

Calmodulin, affinity capture of TAP-tagged complexes, 45–47
Capture array, principles, 153–154
Cold Spring Harbor Laboratory Proteomics Course, 7
Competent cell, preparation, 215–216
Coomasie Blue, staining of two-dimensional gel, 30

Desalting, peptide mixtures with reversed-phase trapping, 211–213
Dipeptides, masses, 201
DNA, quantification, 217

Electrospray ionization. *See* Liquid chromatography-tandem mass spectrometry; Nanoelectrospray ionization-tandem mass spectrometry

Functional proteomics
cloning. *See* High-throughput cloning
overview, 6–7,143
protein microarray. *See* Nucleic acid programmable protein array
synergistic approaches, 143–144

Gel electrophoresis. *See* Two-dimensional gel electrophoresis
Global Proteome Machine (GPM)
database search, 122–126
MudPIT database search, 126
preliminary analysis of database search output, 126–129
program installation on computer, 123
XML format conversion, 121–122

Glycosylation, Pro-Q Emerald staining in two-dimensional gels, 29–30
GPM. *See* Global Proteome Machine

High-performance liquid chromatography (HPLC) *See* Liquid chromatography-tandem mass spectrometry; Multidimensional protein identification technology
High-throughput cloning
clone repository rationale, 144–145
expression construct assembly with Gateway LR reaction, 147–149
recombinatorial cloning, 145–147
transformation of bacteria, 149–152, 215–216
HPLC. *See* High-performance liquid chromatography

IEF. *See* Isoelectric focusing
IgG. *See* Immunoglobulin G
IMAC. *See* Immobilized metal affinity chromatography
Immobilized metal affinity chromatography (IMAC), enrichment of phosphopeptides for mass spectrometry
column
conditioning, 89
construction, 86–88
preprocessing, 89
esterification, 86
liquid chromatography-tandem mass spectrometry, 92
overview, 83–84
reversed-phase capture column preparation, 88–89
sample loading and elution, 90–91
sample preparation, 84–86
trypsin digestion, 85–86
Immobilized pH gradient. *See* Two-dimensional gel electrophoresis
Immunoglobulin G (IgG), affinity capture of TAP-tagged complexes, 43–45
Interactome, 6
IPG strip. *See* Two-dimensional gel electrophoresis

Isobaric tag for relative and absolute
    quantification. *See* iTRAQ
Isoelectric focusing (IEF). *See* iTRAQ; Two-dimensional gel electrophoresis
iTRAQ
    experimental description, 106
    isoelectric focusing of peptides, 111–115
    isotope labeling limitations, 103
    overview, 104–106
    peptide labeling, 109–110
    peptide preparation from yeast
        cell growth, 107–108
        lysis, 108
        material, 107
        protein precipitation, 108
        protein reduction, alkylation, and
            digestion, 108
    ProteinPilot for peptide/protein identification
        data file loading, 130
        database search, 130–132
        overview, 129–130
        results screen, 132–135
    reversed-phase chromatography of peptides,
        110–111

LC-MS/MS. *See* Liquid chromatography-tandem
    mass spectrometry
Liquid chromatography-tandem mass
    spectrometry (LC-MS/MS)
    analysis of peptide tandem mass spectra
        fragmentation spectrum, 117–119
        Global Proteome Machine analysis
            database search, 122–126
            MudPIT database search, 126
            preliminary analysis of database search
                output, 126–129
            program installation on computer, 123
            XML format conversion, 121–122
        nomenclature for fragmentation ions,
            117, 120
        overview, 120–121
        ProteinPilot for peptide/protein
            identification
            data file loading, 130
            database search, 130–132
            overview, 129–130
            results screen, 132–135
        validation of spectrum match after database
            search, 135–140
    assembly, 78–79
    chromatography conditions, 79–82
    immobilized metal affinity chromatography-
        enriched phosphopeptides, 92
    microcapillary high-performance liquid
        chromatography column fabrication
        packing microcapillary fused-silica capillary
            columns, 74–77
        pulling, 71–74
    MudPIT instrument parameters
        materials, 204–205
        VU6_MudPIT_000.meth, 205
        VU6_MudPIT_015.meth, 206–207
        VU6_MudPIT_050.meth, 207–208
        VU6_MudPIT_070.meth, 208
        VU6_MudPIT_100.meth, 209
    one-dimensional reversed-phase nano-liquid
        chromatography/tandem mass
        spectrometry, 203–204
    principles of protein complex analysis, 69–70

Magnetic bead immunoaffinity purification
    incubation, magnetic separation, and elution,
        52–54
    lysis and extraction, 51–52
    magnetic bead conjugation, 48–50
    overview, 48
MALDI/MS. *See* Matrix-assisted laser desorption
    ionization/mass spectrometry
Mass spectrometry
    amino acid masses
        dipeptide masses, 201
        monoisotopic and immonium ion
            masses, 199
    gel spot analysis. *See* Matrix-assisted laser
        desorption ionization/mass
        spectrometry sample
    isobaric tag for relative and absolute quantification. *See* iTRAQ
    liquid chromatography coupling. *See* Liquid
        chromatography-tandem mass
        spectrometry
    multidimensional protein identification
        technology, 101
    principles, 3–5
Matrix-assisted laser desorption ionization/mass
    spectrometry (MALDI/MS)
    sample preparation from two-dimensional gels
        matrix-assisted laser desorption ionization
            plate spotting, 35–36
        overview, 17
        spot cutting, 33
        tryptic digestion in-gel, 34–35, 193–195
    tandem mass spectrometry of peptides
        calibration solution spotting, 58

iTRAQ isoelectric focusing fraction
analysis, 66
key terms and concepts, 56–57
materials, 58
overview, 55–56
sample plate
calibration in reflector mode, 60–64
loading, 59–60
spot set creation, 58–59
two-dimensional gel digest analysis, 64–65
MudPIT. *See* Multidimensional protein identification technology
Multidimensional protein identification technology (MudPIT)
column preparation
connection for electrospray ionization mass spectrometry, 101
overview, 98
packing, 99
Global Proteome Machine database search, 126
liquid chromatography-tandem mass spectrometry instrument parameters
materials, 204–205
VU6_MudPIT_000.meth, 205
VU6_MudPIT_015.meth, 206–207
VU6_MudPIT_050.meth, 207–208
VU6_MudPIT_070.meth, 208
VU6_MudPIT_100.meth, 209
materials, 98–99
principles, 97
sample loading, 99–101
Multidimensional separation, overview, 95–97

Nanoelectrospray ionization-tandem mass spectrometry
applications, 183
materials, 183–184
nanoelectrospray ionization source assembly, 184–186
running conditions for tandem mass spectrometry demonstration, 186–188
NAPPA. *See* Nucleic acid programmable protein array
Nucleic acid programmable protein array (NAPPA)
abundance-based microarrays, 153–154
amino silane coating of slides, 158–159
bacterial culture on microtiter plates, 159–161
chemistry, 156–157
detection on slides
DNA, 172–173
protein, 170–171
DNA plasmids

biotinylation, precipitation, and arraying, 165–167
isolation, 162–165
function-based arrays, 154–156
protein expression on slides, 168–170
protein–protein interaction analysis
advantages, 175
epitope tags, 175–176
query protein
coexpression on slides, 176–178
detection, 178–180

Phosphoproteins
immobilized metal affinity chromatography enrichment of phosphopeptides
column
conditioning, 89
construction, 86–88
preprocessing, 89
esterification, 86
liquid chromatography-tandem mass spectrometry, 92
overview, 83–84
reversed-phase capture column preparation, 88–89
samples
loading and elution, 90–91
preparation, 84–86
trypsin digestion, 85–86
Pro-Q Diamond staining in two-dimensional gels, 28–29
PicoGreen, DNA detection on NAPPA slides, 172–173
Posttranslational modifications
glycosylation. *See* Glycoproteins
overview, 6
phosphorylation. *See* Phosphoproteins
Pro-Q Diamond, staining of phosphoproteins in two-dimensional gels, 28–29
Pro-Q Emerald, staining of glycoproteins in two-dimensional gels, 29–30
Protein complexes
interactome, 6
mass spectrometry analysis. *See* Liquid chromatography-tandem mass spectrometry
NAPPA analysis. *See* Nucleic acid programmable protein array
purification for mass spectrometry analysis
magnetic bead immunoaffinity purification
incubation, magnetic separation, and elution, 52–54
lysis and extraction, 51–52

Protein complexes (*continued*)
    magnetic bead conjugation, 48–50
    overview, 48
  overview, 37
  tandem affinity purification
    calmodulin affinity capture of TAP-tagged complexes, 45–47
    extraction, 41–42
    immunoglobulin G affinity capture of TAP-tagged complexes, 43–45
    principles, 38
    yeast cell growing and harvesting, 39–40
Protein microarray. *See* Nucleic acid programmable protein array
ProteinPilot, peptide/protein identification
  data file loading, 130
  database search, 130–132
  overview, 129–130
  results screen, 132–135
Proteome
  complexity, 1–2
  dynamics, 2

Reverse-phase protein blot (RPPB), principles, 154
Reversed-phase chromatography. *See* Desalting; Immobilized metal affinity chromatography; iTRAQ; Liquid chromatography-tandem mass spectrometry
RPPB. *See* Reverse-phase protein blot

SDS-polyacrylamide gel electrophoresis. *See* Two-dimensional gel electrophoresis
Silver staining, two-dimensional gels for mass spectrometry analysis, 30–31
SYPRO Ruby, staining of two-dimensional gels, 27–28

Tandem affinity purification (TAP)
  calmodulin affinity capture of TAP-tagged complexes, 45–47
  extraction, 41–42
  immunoglobulin G affinity capture of TAP-tagged complexes, 43–45
  principles, 38
  yeast cell growing and harvesting, 39–40
Tandem mass spectrometry. *See* Liquid chromatography-tandem mass spectrometry; Matrix-assisted laser desorption ionization/mass spectrometry; Nanoelectrospray ionization-tandem mass spectrometry
TAP. *See* Tandem affinity purification
TCA. *See* Trichloroacetic acid
Trichloroacetic acid (TCA), protein precipitation, 197
Trypsin digestion
  in-gel, 34–35, 193–195
  phosphoproteins for immobilized metal affinity chromatography, 85–86
  solution digestion, 189–192
Two-dimensional gel electrophoresis
  contamination prevention, 18
  experimental overview, 18–19
  imaging and analysis, 17, 31–32
  isoelectric focusing
    electrophoresis, 22–24
    equilibration of focused strips, 24–25
    IPG strip rehydration, 20–22
    overview, 15–16
  matrix-assisted laser desorption ionization/mass spectrometry sample preparation
    matrix-assisted laser desorption ionization plate spotting, 35–36
    overview, 17
    spot cutting, 33
    tryptic digestion in-gel, 34–35, 193–195
  principles, 2–3, 13, 16
  sample preparation
    cell disruption approaches, 14
    overview, 13–15
    solubilization solution, 14, 19
    yeast cell lysate preparation, 18–19
  SDS-polyacrylamide gel electrophoresis, 16, 25–27
  staining
    Coomasie Blue staining, 30
    Pro-Q Diamond staining of phosphoproteins, 28–29
    Pro-Q Emerald staining of glycoproteins, 29–30
    sensitivity of stains, 16
    silver staining for mass spectrometry analysis, 30–31
    SYPRO Ruby staining, 27–28

XML, mass spectra conversion, 121–122